Handbook
of Advanced
Troubleshooting

Handbook
of Advanced
Troubleshooting

JOHN D. LENK
Consulting Technical Writer

PRENTICE-HALL, INC., Englewood Cliffs, New Jersey 07632

Library of Congress Cataloging in Publication Data

Lenk, John D. (date)
   Handbook of advanced troubleshooting.

   Includes index.
   1. Electronic apparatus and appliances—Maintenance
and repair. I. Title.
TK7870.2.L46 1983   621.381'028'8   82–16603
ISBN 0-13-372391-7

Editorial/production supervision:
   Ros Herion, Oscar Ocampo,
   Natalie Krivanek, Christine Stengel, and Bette Kurtz
Interior design: *Oscar Ocampo*
Cover design: *George Alon Jaediker*
Manufacturing buyer: *Gordon Osbourne*

Printed in the United States of America

10 9 8 7 6 5 4 3

ISBN 0-13-372391-7

PRENTICE-HALL INTERNATIONAL, INC., *London*
PRENTICE-HALL OF AUSTRALIA PTY. LIMITED, *Sydney*
EDITORA PRENTICE-HALL DO BRASIL, LTDA., *Rio de Janeiro*
PRENTICE-HALL CANADA LTD.,*Toronto*
PRENTICE-HALL OF INDIA PRIVATE LIMITED, *New Delhi*
PRENTICE-HALL OF JAPAN, INC.,*Tokyo*
PRENTICE-HALL OF SOUTHEAST ASIA PTE. LTD., *Singapore*
WHITEHALL BOOKS LIMITED, *Wellington, New Zealand*

To
Irene, the most Wonderful Wife in the World;
our daughter *Karen,*
our grandsons *Brandon* and *Justin,*
and to *Lambie,* our little Wookie

Contents

3   COMMUNICATIONS TROUBLESHOOTING   **96**

5   MICROPROCESSOR-BASED DIGITAL TROUBLESHOOTING   197

6   NTSC SIGNALS USED IN TELEVISION AND VCR TROUBLESHOOTING   272

INDEX   303

Preface

The purpose of this handbook is to provide a simplified, practical, but very advanced system of troubleshooting for the three most common areas of electronic equipment: communications, television, and microprocessor-based systems. Although the troubleshooting techniques found in the service literature for such equipment usually provide an orderly approach for locating system faults, they are not always adequate. There are numerous techniques, procedures, and "tricks" that can be effective in diagnosing, isolating, and locating faults in electronic equipment. Many of these advanced techniques are discussed throughout this handbook.

Another purpose of the handbook is to fill the gap between the theory of circuits or systems, and the practical how-to of troubleshooting. The handbook is not just another review of communications, television, or microprocessor theory, nor an oversimplified introduction to theoretical troubleshooting, but a technician-level book on how to determine the cause of circuit and component faults. The handbook is written primarily for working technicians and field service engineers, but it also provides an invaluable guide to the student technician who is about to face the real world of everyday troubleshooting.

The troubleshooting approaches described here are based on the techniques found in the author's best-selling troubleshooting books: *Handbook of Practical Solid-State Troubleshooting, Handbook of Basic Electronic Troubleshooting, Handbook of Simplified Television Service, Handbook of Practical CB Service, Handbook of Practical Microcomputer Troubleshooting,* and *Complete Guide to Video Cassette Recorder Operation and Servicing.* However, this handbook goes beyond all these books to concentrate on shortcuts designed to pinpoint trouble quickly in specific types of equipment. It is assumed that the reader is already familiar with the basics of communications, television, and microprocessor-based equipment at a level found in the author's books. However, no reference to any book is necessary to understand and use this handbook.

Chapter 1 is devoted to a review of basic troubleshooting techniques (evaluation of symptoms, fault isolation, etc.) with special emphasis on communications, television, and microprocessor systems. Such a review is absolutely essential for those readers not already familiar with the author's Universal Troubleshooting Approach. This approach, well known to the author's worldwide audience, can be applied to any electronic equipment not in use, as well as equipment being developed for the future.

Chapters 2 and 4 are devoted to test equipment used in troubleshooting communications and microprocessor-based equipment, respectively. This is particularly important for microprocessor systems since digital networks can be an "electronic nightmare" when hundreds or thousands of circuits are interconnected, as they are in the simplest of microcomputers. Microprocessor-based system faults are located best by analyzing test results (response to input signals, presence or absence of pulses and signal levels, step-by-step tracing of instructions and data through each step of the program, etc.). Chapters 2 and 4 include a variety of test equipment for advanced service or troubleshooting and cover operating principles or characteristics. The discussions describe how features found on present-day test equipment relate to specific problems in troubleshooting.

Chapter 3 is devoted entirely to troubleshooting communications equipment, including AM, FM, and SSB systems. The first part of the chapter describes a series of tests to check all sections of a typical communications set. The second part of the chapter tells how to use the test results to troubleshoot the various circuits. The sequence of the test and troubleshooting procedures is arranged in step-by-step order to get the best possible shortcut for service of a radio communications set.

Chapter 5 is devoted entirely to troubleshooting microprocessor-based digital equipment. Such equipment can be thought of as an extension of traditional digital logic. However, since microprocessor-based systems are bus-structured, since many of the devices on the buses are complex LSI components, and since the buses form feedback paths, very special techniques must be used for advanced troubleshooting. Such techniques include breaking the data bus, which is the system's main feedback path, to help isolate faults. The chapter starts with a discussion concerning some troubleshooting problems unique to microprocessors and goes on through the use of advanced test equipment, including logic analyzers, to localize problems.

Chapter 6 concentrates on the use of NTSC signals in television troubleshooting and on the NTSC color bar generators that are used to produce such signals for field service work. At one time, the NTSC generator was found only in the laboratory or in television broadcast studios. The expense of an NTSC generator made it impractical for the average television service shop. Today, with inexpensive and portable instruments readily available, there is no excuse for not including an NTSC generator in any service facility. The signals produced by such generators simulate those broadcast by television stations and are absolutely essential for proper service and troubleshooting of modern color sets as well as VCRs.

Many professionals have contributed their talent and knowledge to the preparation of this handbook. The author gratefully acknowledges that the tremendous effort required to make this book such a comprehensive work is impossible for one person, and he wishes to thank all who have contributed directly and indirectly. Special thanks are due to the following: B&K Precision Test Instrument Product Group of Dynascan Corporation; General Electric; Heathkit; Hewlett-Packard; Intel Corporation; Mostek Corporation; Motorola Semiconductor Products, Inc.; Pace Communications; Radio Shack; RCA Corporation, Solid State Division; Society of Motion Picture and Television Engineers; Tektronix, Inc.; and Texas Instruments Incorporated.

The author extends his gratitude to Dave Boelio, Hank Kennedy, John Davis, Jerry Slawney, Art Rittenberg, Matt Fox, Dave Ungerer, and Don Schaefer of Prentice-Hall. Their faith in the author has given him encouragement, and their editorial/marketing expertise has made many of the author's books best-sellers. The author also wishes to thank Mr. Joseph A. Labok of Los Angeles Valley College for his help and encouragement.

John D. Lenk

1
Introduction
to Advanced
Troubleshooting

Troubleshooting can be considered as a step-by-step logical approach to locate and correct any fault in the operation of equipment. In the case of electronic equipment, seven basic functions are required.

First, you must study the equipment using service literature, schematic diagrams, and the like to find out how each circuit works when operating normally. In this way, you will know in boring detail how a piece of equipment should work. If you do not take time to learn what is normal, you will never be able to tell what is abnormal. For example, some video cassette recorders (or television receivers) simply have better pictures than others, even in the presence of poor signals. You could waste hours of precious time (which is money if you are a professional) trying to make the inferior equipment perform like the high-quality instrument if you do not know what "normal" operation is.

Second, you must know the function of, and how to manipulate, all equipment controls and adjustments. It is difficult, if not impossible, to check out any equipment without knowing how to set the controls. Besides, it will make a bad impression on the customer if you cannot find the channel selector, especially on the second service call. Also, as electronic equipment ages, readjustment and realignment of critical circuits are often required.

Third, you must know how to interpret service literature and how to use test equipment. Together with good test equipment that you know how to use, well-written service literature is your best friend.

Fourth, you must be able to apply a systematic, logical procedure to locate problems. Of course, a "logical procedure" for one type of equipment is quite illogical for another. As an example, it is quite illogical to check operation of a phase-locked-loop (PLL) circuit on a TV receiver not so equipped. However, it is quite logical to check the automatic gain control (AGC) on any modern TV receiver.

Fifth, you must be able to analyze logically the information you receive from improperly operating equipment. The information to be analyzed may be in the form of performance, such as the appearance of the picture on a TV set, or may be indications taken from test equipment, such as voltage readings. Either way, it is your analysis of the information that makes for logical, efficient troubleshooting.

Sixth, you must be able to perform complete checkout procedures on equipment that has been repaired. Such checkout may only be a simple operation, such as switching through all channels and checking the picture on a TV set. At the other extreme, checkout can involve running through a diagnostic routine of microprocessor-based equipment using the keyboard and display of a video terminal. Either way, checkout is always required after troubleshooting.

One reason for the checkout is that there may be more than one trouble. For example, an aging part may cause high current to flow through a resistor, resulting in burnout of the resistor. Replacement of the resistor will restore operation. However, only a thorough checkout will reveal the original high-current condition that caused the burnout.

Another reason for after-service checkout is that the repair may have produced a condition that requires readjustment. A classic example of this is where replacement of a part changes the circuit characteristics, such as a new transistor in an RF stage of a TV or communications set, and requires complete realignment of all RF stages.

Seventh, you must be able to use the proper tools to repair the trouble. Some electronic devices require elaborate sets of tools (particularly devices that involve mechanical operation, such as video cassette recorders). As a minimum for troubleshooting any electronic equipment, you must be able to use soldering and desoldering tools, wire cutters, longnose pliers, screwdrivers, and socket wrenches. If you are still at a stage where any of these tools seem unfamiliar, you are not ready for any troubleshooting, much less advanced troubleshooting.

In summary, before starting any troubleshooting job, ask yourself these questions: Have I studied all available service literature to find out how the equipment works (including any special circuits such as automatic color control, or automatic noise limiter)? Can I operate the equipment controls properly? Do I really understand the service literature, and can I use all required test equipment properly? Using the service literature, and/or previous experience on similar equipment, can I plan a logical troubleshooting procedure? Can I analyze logically the results of operating checks, as well as checkout procedures involving test equipment? Using the service literature and/or experience, can I perform complete checkout procedures on the equipment, including realignment, adjustment, and so on, if necessary? Once I have found the trouble, can I use common hand tools to make the repairs? If the answer to any of these questions is "no," you will do well to start studying the rest of this book.

1-1. THE UNIVERSAL TROUBLESHOOTING APPROACH

The troubleshooting functions discussed thus far may be divided into four major steps:

1. *Determine* the trouble symptoms.
2. *Localize* the problem to a functional area.
3. *Isolate* the problem to a circuit.
4. *Locate* the specific problem, probably in a specific part.

The remaining sections of this chapter are devoted to these four steps. Before going into the details of the steps, let us examine what is accomplished by each.

1-1.1 Determining Trouble Symptoms

Determining symptoms means that you must know what the equipment is supposed to do when it is operating normally, and that you must be able to recognize when the normal job is not being done. Everyone knows what a TV receiver is supposed to do, but no one knows how well each set is to perform (and has performed in the past) under all operating conditions (with a given antenna, lead-in, location, etc.).

Most electronic equipment has some operating controls, indicating instruments, or other built-in aids for evaluating performance. For example, a TV receiver has operating controls, a picture tube, and a loudspeaker. Similarly, most communications sets have loudspeakers (to indicate voice reception) and meters to indicate received signal strength (S-meter) and transmitted power (relative RF power output), as well as operating controls such as channel selectors, on–off volume controls, and probably a squelch control. Most digital devices have a keyboard and some form of readout (LED or video CRT). Using the normal and abnormal symptoms produced by these indicators, you must analyze the symptoms to ask the questions: How well is this equipment performing; and where in the equipment could there be trouble that would produce these symptoms?

The "determining symptoms" step does not mean that you charge into the equipment with screwdriver and soldering tool, nor does it mean that test equipment should be used extensively. Instead, it means that you make a visual

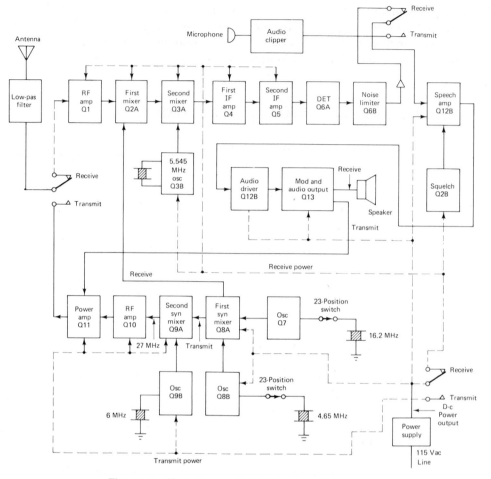

Figure 1-1   Block diagram of typical communications set.

check, noting both normal and abnormal performance indications. It also means that you operate the controls to gain further information.

At the end of the "determining symptoms" step, you definitely know that something is wrong and have a fair idea of what it is, but you probably do not know just what area of the equipment is faulty. This is established in the next step of troubleshooting.

1-1.2 Localizing the Problem to a Functional Unit

Most electronic equipment can be subdivided into units or areas which have a definite purpose or function. The term "function" is used here to denote an operation performed in a specific area of the equipment. For example, the communications set shown in Fig. 1-1 may be divided into receiver, transmitter, power supply, crystal control (frequency synthesizer), transmit-receive/control (switching relay), audio, and low-pass filter, while the TV receiver in Fig. 1-2 may be divided into RF, IF, audio, video, picture tube, and power supply.

To localize the trouble systematically and logically, you must have a knowledge of the functional areas and must correlate all the symptoms previously determined. Thus, you might first determine (by an educated guess) the functional area most likely to cause the indicated symptoms. First, you may consider several technically accurate possibilities as the probable trouble area.

As an oversimplified example of troubleshooting a communications set (Fig. 1-1), if modulation is poor during transmission, and voice is weak during reception, the audio section (Q12 and Q13) is a likely suspect, since it is common to both transmission and reception. On the other hand, if the transmission

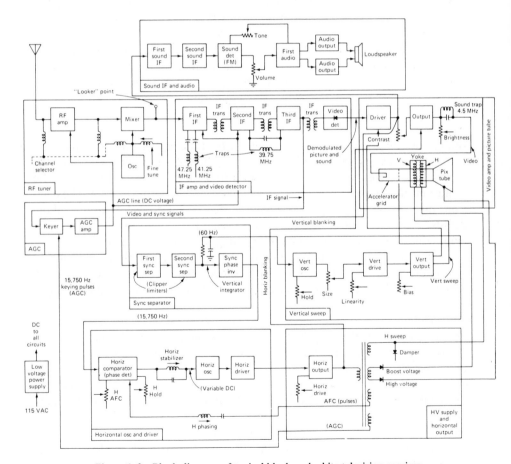

Figure 1-2 Block diagram of typical black-and-white television receiver.

is good but reception is poor, the trouble is probably in the receiver section (Q1 through Q6), since these circuits apply only to reception. In another example of troubleshooting a TV receiver (Fig. 1-2), if both picture and sound are poor, the trouble might be in either the RF or IF stages, since these functional areas are common to both picture and sound reproduction. On the other hand, if the picture is good but the sound is poor, the trouble is probably in the audio stages, since these functional areas apply only to sound.

Use of diagrams.   Troubleshooting involves (or should involve) the extensive use of diagrams. Such diagrams may include a functional *block diagram* and almost always include *schematic diagrams.* The *practical wiring diagram,* once popular in military-style service literature, is not usually found in present-day literature (especially for mass-market or home-entertainment equipment). One reason for this is that parts in most present-day electronic equipment are mounted on printed-circuit (PC) boards. At best, present-day service literature will show the PC board layout, and possibly the wiring from the boards to the external controls, connectors, and indicators.

The block diagram (such as shown in Figs. 1-1 and 1-2) illustrates the functional relationship of all circuits in the equipment, and is thus the most logical source of information for trouble localization. Unfortunately, not all service literature is provided with a block diagram. It may be necessary to use the schematic diagram.

The schematic diagram (such as those shown in remaining chapters) shows the functional relationship of all parts in the equipment. Such parts include all transistors, integrated circuits (ICs), diodes, and so on. Generally, the schematic presents too much information (not directly related to the specific symptoms noted) to be of maximum value during the localizing step. The decisions made regarding the probable trouble area may become lost among all the details. However, the schematic is very useful in later stages of the total troubleshooting effort, or when a block diagram is not available.

In comparing the block diagram and the schematic during the localizing step, note that each transistor or IC shown on the schematic is usually represented as a block on the block diagram. This relationship is typical of most service literature.

The physical relationship of parts is often given on *component location diagrams* (also called *parts placement* or *parts identification diagrams*). These location or placement diagrams rarely show or identify all parts, as do true military-style practical wiring diagrams. Instead, the parts location diagrams concentrate on identification of major parts such as transistors, ICs, and adjustment controls. For this reason, location diagrams are least useful in localizing trouble. Instead, the location diagrams are most useful in locating specific parts during other phases of troubleshooting.

To sum up, it is logical to use a block diagram instead of a schematic or location diagram when you want to make a good guess as to the area of probable trouble. The use of a block diagram also permits you to use a troubleshooting technique known as *bracketing* (discussed in Sec. 1-4). If the block diagram includes major test points, as it may in some well-prepared service literature, the block will also permit you to use test equipment as aids in narrowing down the probable cause of trouble. However, test equipment is used more extensively during the isolation step of troubleshooting.

1-1.3 Isolating the Problem to a Circuit

After the trouble is localized to a single functional area, the next step is to isolate the problem to a circuit in the faulty area. To do this, concentrate on those circuits in the area that could cause the trouble and ignore the remaining circuits.

The isolating step involves the use of test equipment such as meters, oscilloscopes, and signal generators for *signal tracing* and *signal substitution* in the suspected faulty area. By making educated estimates and properly using the applicable diagrams, bracketing techniques, signal tracing, and signal substitution, you can systematically and logically isolate the trouble to a single defective circuit.

Repair techniques or tools to make necessary repairs to the equipment are not used until after the specific problem is located and verified. That is, you still do not charge into the set with soldering tools and pickaxe at this point. Instead, you are now trying to isolate the problem to a specific defective circuit so that, once the problem is located, it can be repaired.

1-1.4 Locating the Specific Problem

Not only does this troubleshooting step involve locating the specific problem, it also includes a final analysis, or review, of the complete procedure, as well as the use of repair techniques to remedy the problem once it has been located. This final analysis permits you to determine whether some other malfunction caused the part to be faulty or whether the part located is the actual cause of trouble.

Inspection by using the senses—sight, smell, hearing, and touch—is very useful in trying to locate the problem. This inspection is usually performed first, in order to gather information that may lead more quickly to the defective part. (The inspection is often referred to as a "visual inspection" in service literature, although it involves all the senses.) Among other things to look for during "visual inspection" are burned, charred, or overheated parts, arcing in the circuit, and burned-out parts.

In equipment where access to the circuitry is relatively easy, a rapid visual inspection should be performed first. Then the active device—transistor or IC—can be checked. A visual inspection is always recommended as the first step in all equipment. A possible exception is equipment in which access to the majority of circuit parts is very difficult, but where certain parts (usually active devices such as transistors and ICs) can be easily removed and tested (or replaced).

The next step in locating the specific problem in typical nondigital equipment is the use of an oscilloscope to observe *waveforms* and a meter to measure *voltages*. As discussed in Chapter 5, digital troubleshooting requires specialized test equipment. Similarly, the oscilloscope is not used as extensively in communications service as in other fields (such as television service) because most of the waveforms in communications are RF, IF, and audio signals rather than pulses (as in TV and digital circuits). However, an oscilloscope should be used to measure the modulation envelope of a communications set, or as a substitute meter to measure voltages. Of course, a conventional meter is best when making *resistance* and *continuity* checks to pinpoint a defective part. After the trouble is located, you should make a final analysis of the complete troubleshooting procedure to verify the trouble. Then you can repair the trouble and check out the equipment for proper operation.

Note that in most present-day service literature the voltages (and possibly the resistances) are often given on the schematic diagram, but this information may also appear in chart form (following the military style). No matter what information is given, and what form it may take, you must be able to use test equipment to make the measurements. For that reason, the function and use of test equipment during troubleshooting is discussed frequently throughout this handbook.

1-1.5 Developing a Systematic and Logical Troubleshooting Procedure

The development of a systematic and logical troubleshooting procedure requires:

A logical approach to the problem
Knowledge of the equipment
The interpretation of test information
The use of information gained in each step

Some technicians feel that a knowledge of the equipment involves remembering past failures as well as such things as the location of all test points, all adjustment procedures, and so on. This approach may be good in troubleshooting only one type of equipment, but it has little value in developing a basic troubleshooting procedure.

It is true that recalling past equipment failures may be helpful, but you should not expect that the same trouble will be the cause of a given symptom in every case. In any electronic equipment, many trouble areas may show approximately the same symptom indications.

Also, you should never rely only on your memory of adjustment procedures, test point locations, and so on, in dealing with any troubleshooting problem. This is one of the functions of service literature containing diagrams and information on the equipment. The remaining chapters discuss the specific use of, and the type of information to be found in, service literature. The important point for you to learn is to be a systematic, logical troubleshooter, not a memory expert.

1-1.6 Relationship Among Troubleshooting Steps

Thus far, we have established the overall troubleshooting approach. Now let us make sure that you understand how each troubleshooting step fits together with the others by analyzing their relationships.

The first step, determining the symptoms, requires the use of the senses, the observation of equipment performance, previous knowledge of equipment operation, the manipulation of operating controls, and possibly the recording of notes. Determining the symptoms presupposes the ability to recognize inappropriate indications, to operate the controls properly, and to note the effect that the controls have on trouble symptoms.

The second step, localizing the problem to a functional area, depends on the information gained in the first step, plus the use of a functional block diagram (or possibly the schematic), and reasoning. During the second step, ask yourself the question: What functional area could cause the indicated symptoms? Then bracketing, or narrowing down the probable defect to a single functional area, is used together with the test equipment to pinpoint the faulty function. The observations in this step depend on noting the indications of the testing devices used to localize the trouble.

The third step, isolating the problem to a circuit, uses all the information gathered up to this time. The main difference between this step and the second step is that now schematics are used instead of block diagrams, and test equipment is used extensively.

In the fourth step, locating the problem, all the findings are reviewed and verified to ensure that the suspected part is the cause of failure. This final step also includes the necessary repair (replacement of defective parts, etc.) as well as a final checkout.

Example of the relationship among troubleshooting steps.   To make sure that you understand the relationships of the troubleshooting steps, let us consider an example. Assume that you are troubleshooting a piece of equipment, you are well into the fourth step ("locate"), and you find nothing wrong with the circuit. That is, all waveforms, voltage measurements, and resistance measurements are normal. What is your next step?

You might assume that nothing is wrong, that the problem is "customer or operator trouble." This is poor judgment. First, there must be something wrong with the equipment since some abnormal symptoms were recognized before you got to the locate step. Never assume anything when troubleshooting. Either the equipment is working properly, or it is not working properly. Either observations and measurements are made, or they are not made. You must draw the right conclusions from the observations, measurements, and other factual evidence, or you must repeat the troubleshooting procedure.

Repeating the troubleshooting procedure. Some technicians new to service work assume that repeating the troubleshooting procedures means starting all over from the first step. Some service literature recommends this since it is possible for anyone, even an experienced technician, to make mistakes. When performed logically and systematically, the troubleshooting procedure keeps mistakes to a minimum. However, voltage and resistance measurements may be interpreted erroneously, waveform observations or bracketing may be performed incorrectly, or many other mistakes may occur through simple oversight.

In spite of such recommendations by other service literature writers, this author contends that "repeat the troubleshooting procedure" means *retrace your steps,* one at a time, until you find the place where you went wrong. Perhaps a previous voltage or resistance measurement was interpreted erroneously in the locate step, or perhaps a waveform observation or bracketing step was incorrectly performed in the isolate step. The cause must be logically and systematically determined by taking a return path to the point at which you went astray.

1-2. APPLYING THE TROUBLESHOOTING APPROACH TO PRACTICAL SITUATIONS

Now that we have established a basic troubleshooting approach, let us discuss how the approach can be applied to the specifics of certain equipment. The remainder of this chapter is devoted to examples of how the approach can be used to troubleshoot communications equipment and TV receivers. These examples are generalized. Detailed descriptions of communications and TV troubleshooting are given in Chapters 2, 3, and 6.

1-3. TROUBLE SYMPTOMS

It is not practical to list all symptoms of troubles that may occur in all communications and TV sets. However, the lists in Figs. 1-3 and 1-4 cover most

POWER SUPPLY
    Set dead, no light in channel indicator or meter
    No transmission, no reception on any channel

RECEIVER (Q1 THROUGH Q6)
    No reception on any channel, transmission normal
    Reception poor, transmission good.

TRANSMITTER (Q9, Q10, Q11)
    No transmission on any channel, reception normal
    Transmission poor, reception good

AUDIO AND MODULATION CIRCUITS (Q12, Q13)
    No modulation, but carrier present (on RF output meter)
    No sound in speaker with volume control full on

AUDIO CLIPPER
    Poor modulation, sound good on reception

FREQUENCY SYNTHESIZER (Q7, Q8)
    Transmission and reception abnormal on certain channels.
    Transmission off-frequency

CONTROL CIRCUITS (SWITCHING RELAY)
    No transmission with push-to-talk switch pressed

LOW-PASS FILTER
    Poor reception and transmission, carrier normal on output meter

Figure 1-3   Typical communications set trouble symptoms.

troubles for a communications set and a black-and-white television set, respectively. The troubles have been grouped into functional areas (or circuits) of the equipment. These areas or circuit groups correspond to those of the block diagrams (Fig. 1-1 for the communications set, Fig. 1-2 for the TV receiver).

Some of the symptoms listed point to only one area of the equipment as a *probable* cause of trouble. For example, in troubleshooting a communications set, if there is no reception on any channel, but transmission is normal on all channels, the trouble is *probably* in the receiver circuits. On the other hand, if there is some problem in both transmission and reception, the trouble *could be* in the low-pass filter, the audio and modulator circuits, the relay switching circuits, or possibly in the power supply. As another example, in troubleshooting a TV receiver, if there is no vertical sweep (the picture-tube screen shows only a horizontal line) but other receiver functions are normal (good sound), the trouble is *probably* in the vertical sweep circuits. On the other hand, if both sound and picture are weak or poor, the trouble *could be* in the RF tuner, in the IF and video detector, or possibly in the driver of the video amplifier and picture-tube circuits.

In the discussions of the localize, isolate, and locate steps, we shall give examples of how the symptoms can be used as the first step in pinpointing trou-

LOW VOLTAGE POWER SUPPLY
    No sound and no picture raster
    No sound, no picture raster, transformer buzz
    Distorted sound and no raster
    Picture pulling and excessive vertical height
HIGH VOLTAGE POWER SUPPLY AND HORIZONTAL OUTPUT
    Dark screen
    Picture overscan
    Narrow picture
    Foldback or foldover
    Nonlinear horizontal display
HORIZONTAL OSCILLATOR
    Dark screen
    Narrow picture
    Horizontal pulling (loss of sync)
    Horizontal distortion
VERTICAL SWEEP
    No vertical sweep
    Insufficient height
    Vertical sync problems
    Vertical distortion (nonlinearity)
    Line splitting
SYNC SEPARATOR
    No sync (horizontal or vertical)
    No vertical sync
    No horizontal sync
    Picture pulling
RF TUNER
    No picture, no sound
    Poor picture, poor sound
    Hum bars or hum distortion
    Picture smearing and sound separated from picture
    Ghosts
    Picture pulling
    Intermittent problems
    UHF tuner problems
IF AMPLIFIER AND VIDEO DETECTOR
    No picture, no sound
    Poor picture, poor sound
    Hum bars or hum distortion
    Picture smearing, pulling or overloading
    Intermittent problems
VIDEO AMP AND PICTURE TUBE
    No raster (dark screen)
    No picture, no sound
    No picture, sound normal
    Contrast problems
    Sound in picture
    Poor picture quality
    Retrace lines in picture
    Intermittent problems
AGC
    No picture, no sound
    Poor picture
SOUND IF AND AUDIO
    No sound
    Poor or weak sound

Figure 1-4   Typical black-and-white television trouble symptoms.

ble to an area, to a circuit within the area, and finally to a part within the circuit. Before going into these steps, here are some notes regarding symptoms.

1-3.1 Determining Trouble Symptoms

It is not difficult to realize that there definitely is trouble when electronic equipment will not operate. For example, there obviously is trouble when a TV receiver is plugged in and turned on and there is no picture, no sound, and no pilot light. A different problem arises when the equipment is still operating but is not doing its job properly. Using the same TV receiver, assume that the picture and sound are present but that the picture is weak and that there is a buzz in the sound.

Another problem in determining trouble symptoms is improper use of the equipment by the operator. In complex electronic equipment, operators are usually trained and checked out on the equipment. The opposite is true of home-entertainment equipment used by the general public. However, no matter what equipment is involved, it is always possible for an operator (or customer) to report a "problem" that is actually a result of improper operation. For these reasons, you must first determine the signs of failure, regardless of the extent of malfunction, caused by either the equipment or the customer. This means that you must know how the equipment operates normally, and how to operate the equipment controls.

1-3.2 Recognizing Trouble Symptoms

Symptom recognition is the art of identifying *normal* and *abnormal* signs of operation in electronic equipment. A trouble symptom is an undesired *change* in equipment performance or a deviation from the standard. For example, in troubleshooting a communications set, the RF output indicator (typically a panel meter) should show some RF output when the push-to-talk button is pressed. If there is no RF indication, you should recognize this as a trouble symptom, because it does not correspond to the normal, expected performance.

Now assume that the same communications set has poor reception, perhaps due to bad signal conditions in the area, or a defective antenna. If the receiver circuits of the set do not have sufficient gain to produce good reception under these conditions, you could mistake this for a trouble symptom, unless you are really familiar with the set. Poor reception (for this particular set operating under these conditions) is not abnormal operation, nor is it an undesired change.

The same thinking can be applied to recognizing trouble symptoms in television service. For example, the normal television picture is a clear, properly contrasted representation of an actual scene. The picture should be centered within the vertical and horizontal boundaries of the screen. If the picture should suddenly begin to roll vertically, you should recognize this as a trouble symptom because it does not correspond to the normal performance that is expected. Now assume that the picture is weak, say due to a poor broadcast signal in the area or a defective antenna. If the RF and IF stages of the receiver do not have sufficient gain to produce a good picture under these conditions, you could mistake this for a trouble symptom. A poor picture (for this particular model of TV operating under these conditions) is not abnormal operation, nor is it an undesired change. Thus, it is not a true trouble symptom and should be so recognized.

1-3.3 Equipment Failure Versus Degraded Performance

Equipment failure means that either the entire equipment or some functional part of the equipment is not operating properly. For example, the total absence of any received signal on a communications set, or the total absence of a picture on the screen of a TV receiver (when all controls are properly set), are forms of equipment failure, even though there may be sound (background noise) from the loudspeaker. Degraded performance is present whenever the equipment is working but is not presenting normal performance. For example, the presence of hum in the loudspeaker is degraded performance, since the equipment has not yet failed but the performance is abnormal. It should be noted that digital equipment does not usually show degraded performance. Instead, digital equipment usually fails completely. However, there can be intermittent digital equipment failures.

1-3.4 Evaluation of Symptoms

Symptom evaluation is the process of obtaining more detailed descriptions of the trouble symptoms. The recognition of the original trouble may not in itself provide enough information to decide on the probable cause or causes of the trouble, because many faults produce similar trouble symptoms.

To evaluate a trouble symptom, it is generally necessary to operate the controls associated with the symptom and apply your knowledge of electronic circuits, supplemented with information gained from the service literature. Of course, the mere adjustment of operating controls is not the complete story of symptom evaluation. However, the discovery of an incorrect setting can be considered a part of the overall symptom evaluation process.

1-3.5 Examples of Evaluating Symptoms

When there is no sound of any kind coming from the loudspeaker of a communications set, there obviously is trouble. The trouble could be caused by a shorted transistor, burned-out diode, defective capacitor, or any one of the several hundred components in the RF, IF, and audio circuits (assuming that power is applied, the set is switched on, and the squelch is properly set). However, the same symptom may be produced when the RF gain is turned down (minimum sensitivity). Think of all the time you may save by checking the operating controls first, before you charge into the set with tools and test equipment. Similarly, when the screen of a TV receiver is not on (no raster), there obviously is trouble. Assuming that power is on, the trouble could be caused by the brightness control being turned down. However, the same symptom can be produced by a burned-out picture tube or a failure of the high-voltage power supply, among many other possible causes.

To do a truly first-rate job of determining trouble symptoms, you must have a complete and thorough knowledge of the normal operating characteristics of the equipment. Your knowledge helps you to decide whether the equipment is doing the job for which it was designed. In most service literature, this is more properly classified as "knowing your equipment."

In addition to knowing the equipment, you must be able to operate all the controls properly in order to determine the symptom; to decide on *normal* or *abnormal* performance. If the trouble is cleared up by manipulating the controls, your analysis may or may not stop at this point. Through your knowledge of the equipment, you should be able to understand why a specific control adjustment removed the apparent trouble.

1-4. LOCALIZING TROUBLE

Localizing trouble means that you must determine which of the major functional areas in the equipment are actually at fault. This is done by systematically checking each area selected until the faulty one is found. If none of the functional areas on your list shows improper performance, you must take a return path and recheck the symptom information (and observe more information, if possible). Several circuits could be causing the trouble, and the localize step narrows the list to those in one functional area, as indicated by a particular block of the block diagram.

The problem of trouble localization is simplified when a block diagram and a list of trouble symptoms are available for the equipment being serviced. Figure 1-3 lists typical trouble symptoms for a communications set, and is based on the block diagram of Fig. 1-1. Figure 1-4 lists trouble symptoms for a TV receiver, and is based on the block diagram of Fig. 1-2. Keep in mind that these illustrations apply to "typical" or composite equipment. However, the general arrangement shown in Figs. 1-1 through 1-4 can be applied to many similar sets. Thus, the illustrations serve as a universal starting point for trouble localization in communications and television equipment.

1-4.1 Bracketing Technique

The basic bracketing technique makes use of the block diagram or schematic to localize the trouble to a functional area. Bracketing (sometimes known as the *good-input/bad-output* technique) provides a means of narrowing down the trouble to a circuit group and then to a faulty circuit. Symptom analysis and/or signal-tracing tests are used in conjunction with, or are a part of, bracketing.

Bracketing starts by placing brackets (at the good input and the bad output) on the block or schematic. Bracketing can be done mentally, or it can be physically marked with a pencil, whichever is most effective for you. No matter what system is used, with the brackets properly positioned, you know that the trouble exists somewhere between the two brackets.

The technique involves moving the brackets, one at a time (either the good input or the bad output), and then making tests to find if the trouble is within the newly bracketed area. The process continues until the brackets localize a circuit group.

The most important factor in bracketing is to find where the brackets should be moved in the elimination process. This is determined from your deductions based on your knowledge of the set and on the symptoms. All moves of the brackets should be aimed at localizing the trouble with a minimum of tests.

1-4.2 Examples of Bracketing

Bracketing may be used with or without actual measurement of voltages or signals. That is, sometimes localization can be made on the basis of symptom evaluation alone. In practical troubleshooting, both symptom evaluation and tests are usually required, often simultaneously. The following examples show how the technique is used in both cases.

Assume that you are servicing a communications set and that you find a "no-reception, no-transmission" symptom. That is, the power is applied, the set is turned on, the pilot lamps are on, but there is no RF indication on the panel meter during transmission and no signal strength indication during reception. The power supply is a logical suspect as the faulty circuit group.

You place a good-input bracket at the 115-V input, and a bad-output bracket at the d-c output, as shown in Fig. 1-5. To confirm the symptom, you measure both the d-c output voltage (or voltages) and the a-c input voltage. If the input is normal but one or more of the output voltages is absent or abnormal, you have localized the trouble to the power-supply circuits. The next step is to isolate the trouble to a specific circuit in the power supply, as discussed in Sec. 1-5.

From a practical troubleshooting standpoint, it is possible that the power-supply output voltages are normal, but you still have a "no-reception and no-

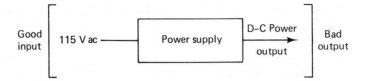

Figure 1-5 Example of bracketing for "set dead" symptom.

transmission" symptom. For example, the lines carrying the d-c voltages to other circuit groups could be open, or the switching relay circuits could be malfunctioning to interrupt the d-c voltages. This can be checked by measuring the voltages at the circuit end of the lines as well as at the power-supply end.

Also, it is possible that a failure in another circuit could cause the power-supply output voltage to be abnormal. For example, if there is a short in one of the circuits on the d-c supply line, the d-c output voltage will be low. Of course, this will show up as an abnormal measurement and will be tracked down during the "isolate" step of troubleshooting.

As another example of bracketing, now assume that the "no-reception, no-transmission" condition still exists, but that the symptoms are somewhat different. Now, there is an RF indication on the panel meter during transmission, and a signal strength indication during reception, but there is no sound in the loudspeaker. You could start by placing a good-input bracket at the input to the audio and modulation circuits, and a bad-output bracket at the output of these circuits (loudspeaker and/or modulation transformer), as shown in Fig. 1-6. However, from a practical standpoint, your first move should be adjustment of the volume and squelch controls.

If the trouble is not cleared by adjustment of the controls, confirm the good-input bracket by monitoring the signal at the audio and modulator circuit

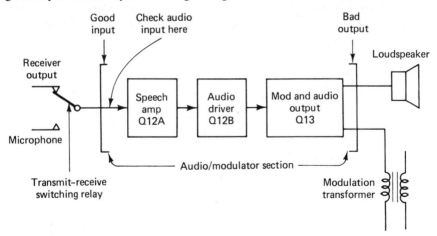

Figure 1-6 Example of bracketing for "no-reception, no-transmission, good S/RF-meter indications" symptom.

input (during reception there should be audio at this point). Make this check at the input to the audio, as shown in Fig. 1-6. Possibly the line between the receiver and audio circuits is open, or perhaps the switching relay contacts are defective. It is also possible that the line is partially shorted. (A completely shorted line would probably cause failure of the receiver circuits, and could result in a lack of signal indication on the panel meter.)

If there are audio signals at the input of the audio and modulator circuits but there is no sound on the loudspeaker (even with adjustment of the volume and squelch controls), you have localized the trouble to the audio and modulator circuits. The next step is to isolate the trouble to a specific circuit in the audio and modulator group, as discussed in Sec. 1-5.

As another example of bracketing, assume that you are servicing a television receiver, and there is no vertical sync. That is, all receiver functions are normal but the picture rolls vertically. You could start by placing a good-input bracket at the input to the vertical sweep circuits and a bad-output bracket at the vertical coils of the deflection yoke, as shown in Fig. 1-7. However, from a practical standpoint, your first move should be adjustment of the vertical hold control.

If the trouble is not cleared by adjustment of the vertical hold control, confirm the good-input bracket by measuring the input sync pulses. Make this measurement at the input to the vertical oscillator rather than at the sync separator, as shown in Fig. 1-7. It is possible that the line between the sync separator and vertical oscillator is open or partially shorted. (A completely shorted line would probably cause failure of the sync separator circuits and would affect the horizontal sweep circuits as well.)

If the vertical sync pulses are normal at the input of the vertical oscillator but there is no vertical sync (even with adjustment of vertical hold), you have localized the trouble to the vertical sweep circuits. The next step is to isolate the trouble to a specific circuit in the vertical sweep group, as discussed in Sec. 1-5.

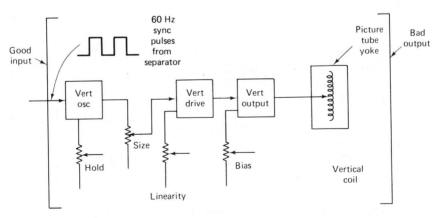

Figure 1-7   Example of bracketing for "no vertical sync" symptom.

1-4.3 Localization with Replaceable Modules

The localization procedure can be modified when the circuits are located on replaceable modules. The trend in present-day equipment is toward the use of replaceable modules, such as PC boards that are either plug-in or require only a few soldered connections. In such equipment it is possible to replace each module or board in turn until the trouble is cleared. For example, if replacement of the audio module restores normal operation, the defect is in the audio module. This conclusion can be confirmed by reinserting the suspected defective module. Although this confirmation process is not a part of theoretical troubleshooting, it is a good practical check, particularly in the case of a plug-in module. Often, a trouble symptom of this sort may be the result of a poor contact between the plug-in module and the chassis connector or receptacle.

Some service literature recommends that tests be made before all modules are arbitrarily replaced, usually because the modules are not necessarily arranged according to functional area. Thus, there is no direct relationship between the trouble symptom and the modules. In such cases, always follow the service literature recommendations. Of course, if modules are not readily available in the field, you must make tests to localize the trouble to a module (so that you can order the right module, for example). Also, operation controls and connectors are not usually found on replaceable modules, so they must be tested separately.

1-4.4 Which Circuit Groups to Test First

When you have localized trouble to more than one circuit group you must decide which group to test first. Several factors should be considered in making this decision.

As a rule, if you can run a test that eliminates several circuits, or circuit groups, use that test first, before making a test that eliminates only one circuit. This requires an examination of the diagrams (block and/or schematic) and a knowledge of how the set operates. The decision also requires that you apply logic.

Test point accessibility is the next factor to consider. A test point can be a special jack located at an accessible spot (say at the top of the chassis). The jack (or possibly a terminal) is electrically connected (directly or by a switch) to some important operating voltage or signal path. At the other extreme, a test point can be any point where wires join or where parts are connected together.

Another factor (although definitely not the most important) is your past experience and a history of repeated failures. Past experience with identical or similar sets and related trouble symptoms, as well as the probability of failure based on records of repeated failures, should have some bearing on the choice of a first test point. However, all circuit groups related to the trouble symptom should be tested, no matter how much experience you may have had with the equipment. Of course, the experience factor may help you decide which group to test first.

Anyone who has had any practical experience in troubleshooting knows that all the steps of a localization sequence rarely proceed in textbook fashion. Just as true is the fact that many troubles listed in the service literature may never occur in the equipment you are servicing. These troubles are included in the literature as a guide, and are not meant to be hard and fast rules. In some cases of localizing the trouble, it may be necessary to modify your trouble-shooting procedure. The physical arrangement of the equipment may pose special troubleshooting problems. Also, special knowledge gained from experience with similar sets may simplify the task of localizing the trouble.

1-4.5 Universal Trouble Localization

In the following paragraphs, we describe universal trouble localization procedures for a typical communications set and a black-and-white TV receiver. The procedures are based on the assumption that the circuit arrangements are as shown in Figs. 1-1 and 1-2; thus, it is possible to group the troubles as shown in Figs. 1-3 and 1-4.

Communications set trouble localization (Figs. 1-1 and 1-3).   If the set is completely "dead" (no panel lights, no transmission, no reception), check the input and output of the power supply. Also check the power-supply fuse (if any). If one or more power-supply output voltages are absent or abnormal, you have localized the problem to the power-supply circuits. If the power-supply output voltages are normal, check for proper distribution of voltages at the remaining circuits (receiver, transmitter, etc.). If the receiver voltages are normal, but not the transmitter (or vice versa), check the relay switching circuits.

If there is no reception, or reception is poor, but transmission is normal (with all controls properly set), check for a signal at the receiver output with the receiver tuned to an active channel. The volume control is generally a convenient test point, as shown in Fig. 1-8. The audio signal can be monitored (traced) at this point (to check operation of the receiver circuits Q1 through Q6). As an alternative, an audio signal can be injected at the volume control (to check operation of the audio and modulation section Q12 through Q13), as shown in Fig. 1-9. The same tests can be made at the relay contacts that switch the audio and modulation circuit input between the receiver output and audio clipper. Either way, if an audio signal is present at the volume control or relay contacts, the receiver circuits are cleared. Using the alternative test, if the audio signal passes to the speaker, the audio and modulator circuits are cleared.

Instead of signal tracing through the receiver circuits, the receiver can also be checked by injecting a modulated RF signal at the antenna jack (the RF signal must be modulated by an audio tone) and listening for the modulated tone in the speaker, as shown in Fig. 1-10. Since this symptom may be caused

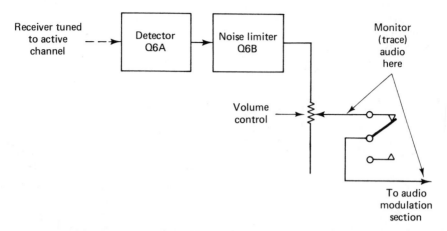

Figure 1-8   Monitoring audio output of receiver to trace a "no-reception or poor-reception" symptom.

by a defect in the switching control circuits, inject the modulated RF signal on both sides of the relay contacts as shown.

To eliminate the AVC circuits as trouble suspects, apply a fixed d-c voltage to the AVC line and check operation of the receiver. (This is known as *clamping* the AVC line.) If operation is normal with the AVC line clamped but not when the clamp is removed, you have localized trouble to the AVC circuit.

It should be noted that AVC circuit problems are often difficult to localize, because an AVC circuit uses feedback. For example, if the IF amplifiers are defective, the detector Q6A and the AVC circuit will not receive proper IF signals. (The AVC voltage is developed by the detector Q6A circuit.) In turn, the lack of proper AVC voltage may cause the IF amplifiers to operate improperly. Conversely, if the AVC circuits are defective, the IF amplifiers will not receive proper AVC voltages and will not deliver a proper IF signal to the AVC circuits.

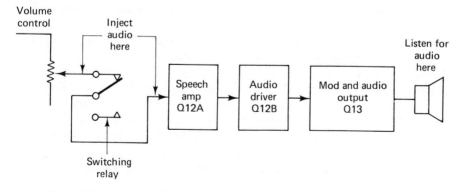

Figure 1-9   Injecting audio at receiver output to clear audio/modulation section.

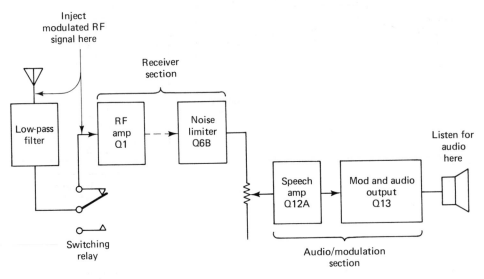

Figure 1-10   Injecting modulated RF at the receiver input to check receiver and audio/modulation sections.

As a rule of thumb, note that if clamping the AVC line eliminates the "no-reception" symptom, the trouble is probably localized to the AVC circuits.

If there is no transmission, or transmission is poor but reception is normal, check for an RF indication on the front-panel meter (with the push-to-talk button pressed). If the RF indication is absent or abnormal (very low), you have localized trouble to the transmitter circuits (Q9, Q10, and Q11).

This symptom may also be caused by a defect in the switching control circuits. Check to see that the relay operates when the push-to-talk button is pressed, and that all sets of contacts switch from receive to transmit (antenna from receiver input to transmitter output, power from receiver to transmitter, audio input from receiver to microphone or audio clipper input, and the audio output to speaker is disabled). If any one of these functions is not normal, trouble is localized to the switching control circuits.

If there is no modulation on transmission, and no background noise in the speaker during reception (with the squelch properly set), the circuits can be checked quickly by injecting an audio signal at the input (volume control or corresponding relay contacts) and listening for a tone in the speaker, as in the process for a "no-reception or reception poor" symptoms, as shown in Fig. 1-9.

If there is no modulation on transmission but there is background noise in the speaker during reception, the problem is likely to be in the audio clipper circuits. These circuits may be checked by injecting an audio signal at the microphone input and checking at the corresponding relay contacts, as shown in Fig. 1-11.

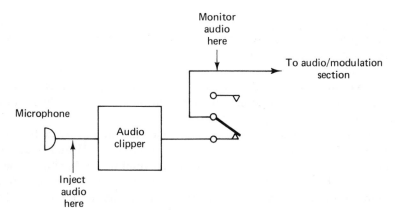

Figure 1-11   Checking audio signals through the audio clipper circuits.

The same symptoms can be produced by a defective microphone; this may be checked by substitution. Finally, it is possible that the symptom is caused by bad relay contacts (at the audio and modulator circuit input).

Note that some communications sets are provided with a modulation indicator. Often this indicator is a panel lamp that glows when modulation is applied (strong modulation produces a bright glow, and weak modulation produces a dull glow). Such an indicator makes it easy to localize modulation problems. For example, if the modulator lamp is not glowing (to indicate modulation) or if the lamp glow remains fixed when modulation is applied but the panel meter indicates proper RF output, the problem is between the microphone and the modulation output circuit.

If transmission and reception are absent or abnormal only on certain channels, the problem is likely to be in the frequency synthesis (crystal control) circuits, or in the channel selector switching circuits. Either way, problems that produce these symptoms are usually easy to localize. However, the problem may not be easy to isolate once you are into the circuits, particularly on those sets where digital control is used for channel selection. Digital troubleshooting is described in Chapters 4 and 5.

If transmission is poor but the RF indication is normal (on the front-panel meter), and reception is poor (weak signal strength indication) but background noise appears normal, the low-pass filter is a likely suspect. For example, the low-pass filter could have shorted capacitors or open coils. (Note that the low-pass filter of Fig. 1-1 is used to prevent harmonics of the signals being transmitted by communications from reaching the antenna.)

The same symptoms may be caused by a defective antenna system. Needless to say, it is possible to spend hours trying to localize problems in a perfectly good communications set if the antenna or lead-in is bad (shorted coax, improper connections, mismatching, etc.). When you are confronted with some mysterious "poor reception and poor transmission" symptom, always try the set on a known good antenna. Also, check the antenna as described in Chapters 2 and 3.

Television trouble localization (Figs. 1–2 and 1–4).　　If there is a "no-sound and no-raster" symptom, start localization by checking the input and output of the low-voltage power supply.

If the raster is present but there is no sound and no picture on the raster, start localization by checking the signal at the output of the IF and video detector and at the output of the driver in the video amplifier and picture-tube circuits. Start at the same points if you have a poor picture, poor sound trouble symptom (or any symptom that points to circuits common to both sound and picture).

If the driver output is abnormal but the IF output is good, you have localized the trouble to the video driver. If the IF output is abnormal, you have localized the trouble to the IF and video detector, RF tuner, or AGC circuits.

To eliminate the AGC circuits as a trouble suspect, apply a fixed d-c voltage to the AGC line, and check the operation of the receiver. (This is known as *clamping* the AGC line.) If operation is normal with the AGC line clamped but not when the clamp is removed, you have localized trouble to the AGC circuits.

It should be noted that AGC circuit problems are often difficult to localize, because a keyed AGC circuit as shown in Fig. 1–2 (and as in millions of receivers) requires two inputs (pulses from the horizontal section and IF signals from the IF section).

For example, if the IF amplifiers are defective, the AGC circuits do not receive proper IF signals and do not produce a proper d-c voltage to the AGC line. In turn, lack of proper AGC voltage can cause the IF amplifiers to operate improperly. Conversely, if the AGC circuits are defective, the IF amplifiers do not receive proper AGC voltages and do not deliver a proper IF signal to the AGC circuits.

As a simple guide, if clamping the AGC line eliminates the problem, the trouble is probably localized to the AGC circuits. In that event, check both inputs (pulses and IF) to the AGC circuits.

If the AGC line is clamped and the IF and video detector output is not normal, check the signal at the IF section input. (This is at the input of the first IF amplifier, which is also the RF tuner output. Many receivers have a readily accessible test point at this input/output junction, often known as the "looker point".)

If the signal at the looker point is abnormal, the trouble is localized to the RF tuner. If the looker point signal is normal but the video detector output is abnormal, you have localized trouble to the IF and video detector section.

1-4.6 Obvious Symptoms

Some trouble symptoms lead you to an obvious localization process. For example, in troubleshooting a TV, if there is a problem on only one channel (in picture, or sound, or both), the RF tuner is the likely suspect. If the problem is one of sound only, with a good picture, the sound IF and audio circuit group is suspect. If there is good sound and a good picture raster but a poor picture, start localization at the video amplifier and picture-tube circuit group.

If horizontal sync is good, together with good sound, but there is a vertical problem (lack of sync, insufficient height, poor linearity, etc.), look for problems in the vertical sweep circuits. If there is good horizontal and vertical sync but the picture is narrow or there is obvious distortion, start with the high-voltage supply and horizontal output circuit group (probably with the horizontal output stage). If both horizontal and vertical sync appear to be abnormal, start by checking the input and both outputs of the sync separator.

1-4.7 Ambiguous Symptoms

The localization process is not always so obvious. For example, in troubleshooting a TV, the horizontal circuits can be a particular problem. If the horizontal oscillator or driver fail completely, there are no pulses to the high-voltage supply and horizontal output circuits. Thus, there is no high voltage to the picture tube (and no picture raster). The identical symptom can be caused by failure of the high-voltage transformer, the high-voltage output stage, or the picture tube itself. (In some receivers, the high-voltage and output circuit group also supplies voltages to the picture tube accelerator grids. If these voltages are absent, there is no raster.) Also, in some receivers, a failure in the video amplifier and picture-tube circuits can cut the picture tube off.

If you are faced with a "no-raster" or "dark-screen" symptom but have sound, there are two logical courses of localization. As a first choice, if convenient, check the high voltage to the picture tube as well as the voltages to all picture-tube elements (accelerator grids, cathode, heater, etc.). As a second logical choice, if it is not convenient to check the picture tube voltages, check the input to the high-voltage and horizontal output group (which is also the output of the horizontal oscillator and driver circuit group). This second choice of localization must also be followed if you find the picture-tube voltages normal. If the signal at the horizontal oscillator and driver output is normal, the high-voltage and horizontal output group is suspect. If the signal is abnormal, the horizontal oscillator and driver is the likely problem area.

Failure in the horizontal circuits can also affect the AGC circuit. If there are no horizontal pulses to the keyer, the d-c output from the AGC circuits is absent. This is a particular problem since failure of the AGC can affect both the

RF tuner and IF circuits, which, as discussed, affect overall performance of the receiver.

When the raster is present, and you have good sound, but there is abnormal horizontal sync (or distortion and/or picture pulling), start with the horizontal oscillator and driver group. Check both the input pulses from the sync separator and the feedback pulses from the horizontal output.

1-5. ISOLATING TROUBLE TO A CIRCUIT

The first two steps (symptoms and localization) of the troubleshooting procedure give you the initial symptom information about the trouble and describe the method of localizing the trouble to the probable faulty circuit group. Both steps involve a minimum of testing. In the isolate step, you do extensive testing to track the trouble to a specific faulty circuit.

1-5.1 Isolating Trouble in IC and Plug-in Equipment

ICs are in common use throughout present-day electronic equipment. For example, in some communications sets, the entire IF circuit or a noise blanker amplifier circuit is replaced by a single IC. In television, the entire IF and video detector circuit group is replaced by an IC. All parts of the circuit group except transformers are included in the IC. The module may be plug-in, but it usually requires some solder connections.

In electronic equipment with ICs and replaceable modules, the trouble can be isolated to the IC or module input and output, but not to the circuits (or individual parts) within the IC. No further isolation is necessary, since parts within the IC cannot be replaced on an individual basis. This same condition is true of some solid-state sets where groups of circuits are mounted on sealed, replaceable boards or cards. Note that not all modules are sealed; many have replaceable parts.

1-5.2 Using Diagrams in the Isolation Process

No matter what physical arrangement is used, the isolation process follows the same reasoning you have already used: the continuous narrowing down of the trouble area by making logical decisions and performing logical tests. Such a process reduces the number of tests that must be performed, thus saving time and reducing the possibility of error.

A block diagram is a convenient tool for the isolation process, since it shows circuits already arranged in circuit groups. Unfortunately, as discussed, you may or may not have a block diagram supplied with your service literature; you must work with a schematic diagram.

With either diagram, if you can recognize circuit groups as well as individual circuits, the isolation process is much easier. For example, if you can subdivide (mentally or otherwise) the schematic diagram of the equipment you are troubleshooting into circuit groups rather than individual circuits, you can isolate the group by a single test at the input or output of the group.

Let us study an example of using diagrams in the isolation process. The block diagram in Fig. 1-1 is arranged into individual circuits, with each block representing a transistor and its related circuit parts. You can arrange these blocks into circuit groups, as is done in Fig. 1-3, where the blocks representing Q1 through Q6 form the receiver circuit group, blocks Q7 and Q8 form the frequency synthesizer, Q9 through Q11 form the transmitter, and Q12/Q13 form the audio and modulation group. Most communications sets have a similar (but not identical) arrangement. Make it a practice to group the circuits mentally on the block or schematic as a first step in isolating trouble.

No matter what diagram you use, or what equipment arrangement is found, you are looking for three major bits of information: the signal path (or paths), the signal form (waveform, amplitude, frequency, etc.), and the operating/adjustment controls in the various circuits along the signal paths. If you know what signals are supposed to go where, and how the signals may be affected by controls, you can isolate trouble quickly in any electronic equipment.

In Fig. 1-1, the receive signal paths are indicated by heavy dashed lines with an arrow; transmit signal paths are shown by a heavy, solid line with arrows. This arrangement is unique to the block of Fig. 1-1. Always study the service literature diagrams for any special notations. Note that both receive and transmit signal paths are shown coming from the synthesizer mixer Q8A and going through the audio and modulation circuits Q12/Q13.

No waveforms are shown in Fig. 1-1. This is standard for most communications set diagrams (both block and schematic) since the signals in communications equipment are typically sine waves (AF, IF, or RF), and the shape or form is not critical (with the possible exception of the audio section). As discussed in Sec. 1-6, waveforms are often found on the schematics of television receivers and similar equipment. In communications equipment, you are interested in amplitude and frequency of the signals. In most communications equipment literature, the amplitude/frequency information is found in the alignment and adjustment instructions rather than on diagrams. Once you have arranged the individual circuits into groups, your next step is to locate all operating and adjustment controls. These controls are always shown on the schematic, but may or may not be shown on block diagrams.

1-5.3 Comparison of Signals or Waveforms

In its simplest form, the isolation step involves comparing the actual signals or waveforms produced along the paths of the circuits against the signals given in the service literature. This is known as *signal tracing*. The isolation step may also involve *injection* or *substitution* of signals normally found along the signal paths. Signal tracing and injection are discussed in Sec. 1–6, as well as throughout the rest of this book. With either technique, you check and compare inputs and outputs of circuit groups and circuits in the signal paths.

In vacuum-tube equipment (yes, it still exists), the input is injected at the grid, and the output signal is traced at the plate (or possibly the cathode). In solid-state equipment, the input signal is injected at the base, and the output signal is traced at the collector (or possibly the emitter). These input/output relationships are shown in Fig. 1–12.

For a circuit group, the input is at the first base (or grid) in the signal path, whereas the output is at the last collector (or plate) in the same path. In any circuit group, the input signal to the group is injected at one point, and then the output signal is obtained at a point several stages farther along the same signal path. To determine the signal-injection and output points of a circuit group, you must find the first circuit of the group in the signal path and the final circuit of the group in the same path.

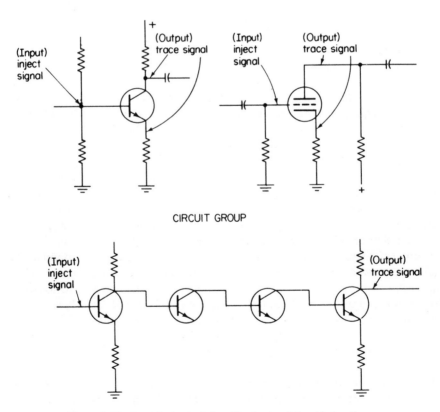

Figure 1-12   Input/output relationships in circuit troubleshooting.

Consider the following example. In the block diagram of Fig. 1-1, the input for the audio and modulator circuit group is at the relay contacts that switch between the audio clipper output and the noise limiter output. The output from the same circuit group is at the loudspeaker during receive, and at the modulation input of the power amplifier Q11 during transmit. Note that the relay contacts are at the input of the audio group and simultaneously at the output of the receiver or clipper groups. Since relay contacts of most communications sets are readily identifiable (and accessible), they form good input/output test points for universal troubleshooting.

Before going on with trouble isolation, keep the following points in mind. The symptoms and related information obtained in the previous steps (symptoms and localization) should not be discarded now or at any time during the troubleshooting procedure. From this information you are able to identify those circuit groups that are probably trouble sources. Also, note that the physical location of the circuit groups within the equipment has no relation to their representation on the diagrams (block or schematic). You must consult part placement diagrams to find physical location.

1-5.4 Signal Tracing Versus Signal Substitution

Both signal tracing and signal substitution (or signal-injection) techniques are used frequently in troubleshooting all types of electronic equipment. The choice between tracing and substitution depends on the test equipment used and the circuit involved. In the case of communications equipment, signal tracing is generally used for the oscillator, frequency synthesizer, and transmitter RF amplifier circuits, because these circuits generate or amplify signals that are readily traced and need not be substituted. Signal substitution (or injection) is generally used for the receiver RF, IF, noise, squelch, and audio circuits.

As discussed in Chapters 2 and 3, signal generators designed specifically for communications use have outputs that simulate signals found in all major signal paths of the receiver. If you have such a generator, signal injection is the logical choice, since you can test all the circuit groups individually (independently of other circuit groups) from antenna to the loudspeaker and/or S/RF-meter. As discussed in Chapter 6, there are signal generators designed specifically for television service. However, it is possible to troubleshoot the receiver circuits of most communications sets, as well as the majority of television circuits with signal tracing alone (and this technique is often recommended by many experienced technicians).

Signal tracing is done by examining the signals at test points with a monitoring device such as a frequency counter, oscilloscope, multimeter, or loudspeaker. In signal tracing, the input probe of the indicating or monitoring device used to trace the signal is moved from point to point, with a signal applied at a fixed point. The applied signal may be generated from an external device, or the normal signal associated with the equipment may be used.

Signal substitution is done by injecting an artificial signal (from a signal generator) into a circuit or circuit group (or the complete equipment) to check performance. In signal injection, the injected signal is moved from point to point, with an indicating or monitoring device remaining fixed at one point. The monitoring can be done with external test equipment or with the equipment indicators (such as the S/RF-meter of a communications set, or the screen of a TV).

Both signal tracing and substitution are often used simultaneously in troubleshooting. For example, in troubleshooting the receiver section of a communications set, it is common practice to inject a modulated RF signal at the antenna and monitor the output with a meter or scope.

1-5.5 Half-Split Technique

The half-split technique is based on the idea of using any test that eliminates the maximum number of circuit groups or circuits simultaneously. This saves both time and effort. Using half-split, you place brackets at good-input and bad-output points in the normal manner, and study the symptoms. Unless the symptoms point definitely to one circuit or circuit group that might be the trouble source, the most logical place to make the first test is at a convenient test point halfway between the brackets.

Example of the half-split technique in communications trouble-shooting. The block diagram in Fig. 1-13 is a simplified version of the receiver circuit group for the communications set shown in Fig. 1-1. Note that in Fig. 1-12 the first and second mixers are combined into one block, as are the first and second IF amplifiers. Figure 1-13 thus shows the signal path of the receive signal from the antenna to the volume control. The brackets placed at the antenna (good-input) and volume control (bad-output) show the trouble being localized to the receiver circuit group. The brackets have been placed at these points as a result of a "no reception on any channel, transmission normal" trouble symptom, or a "reception poor, transmission good" symptom, as shown in Fig. 1-3.

The next phase of troubleshooting is to isolate the trouble to one of the circuit groups (mixers or IF amplifiers) or to one of the individual circuits (RF amplifier, detector, or noise limiter) in the signal path. The selection of test points during this phase depends on their accessibility and the method of troubleshooting (signal tracing or signal injection).

Assuming that test points A, B, C, and D are equally accessible (and that there are no special symptoms that would point to a particular circuit or group), test point C is the most logical point for the first test if signal tracing is used. Test point B is the next most logical choice. When using signal injection, however, test points C and D are the most logical choices. The discussion that follows describes the reasoning for making these choices.

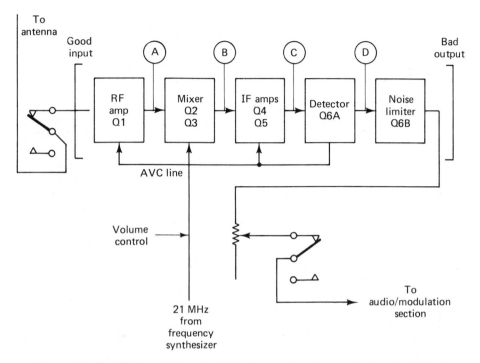

Figure 1-13   Simplified troubleshooting block diagram of receiver section showing linear signal path from antenna to volume control.

Half-split technique using signal tracing.   If signal tracing is used, an RF signal (at the selected channel frequency) modulated by an AF tone is introduced at the antenna. A monitoring device is then connected to monitor the signal at various test points, as shown in Fig. 1-14. The monitoring device may be a meter or oscilloscope with suitable probes or a frequency counter. The meter or scope shows the signal amplitude, but not the frequency. The counter shows the signal frequency, but not the amplitude.

Test point C is a logical choice for a first test.   If C is chosen first and the monitoring display is normal, you have cleared four circuits (RF amplifier, mixer, local oscillator, and IF amplifiers). You have also established the fact that there is a 21-MHz signal from the frequency synthesizer. However, there may still be defective circuits (detector and noise limiter). Note that this process divides the circuits into two groups (known good and possibly bad).

Now assume that the indication at test point C is abnormal (totally absent, low in amplitude, off-frequency, etc.). The bad-output bracket can be moved to test point C, with the good-input bracket remaining at the antenna. The next logical test point is B, because B is near the halfway point between the

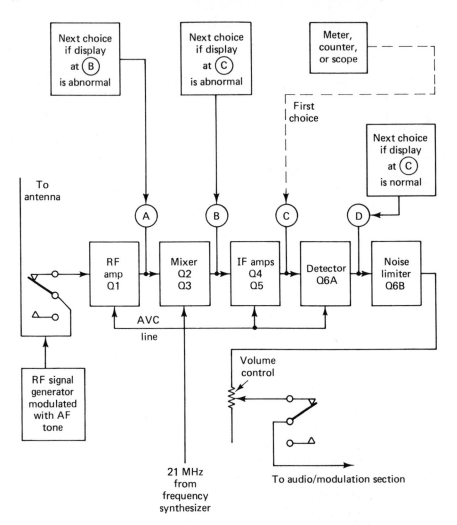

Figure 1-14 Example of half-split technique using signal tracing.

two brackets. (If you choose point A instead of point B, you confirm or deny the possibility of trouble in the RF amplifier *only*.)

If the monitoring indication is abnormal at test point B, the trouble is isolated to the RF amplifier or the mixers. (The term "mixer" used here includes the local oscillator and the 21-MHz signal from the frequency synthesizer.) Additional observation at test point A further isolates the trouble to either circuit (RF amplifier or mixer).

The final step in this half-split process is to monitor the signal at test point A. If there is an abnormal indication at A, the bad-output bracket can be moved to A and the trouble isolated to the RF amplifier. If there is a normal display at A, the good-input bracket can be moved to A and the trouble isolated to the mixer (including the local oscillator).

Now, let us see what happens when a test point other than C is monitored first, using signal tracing.

If you choose test point A for the first test and get a normal indication, the trouble is located somewhere between A and the volume control. This means that the trouble could be in the mixers, IF amplifiers, detector, or noise limiter; you still have many test points to check.

On the other hand, if you get an abnormal signal at A, the trouble is immediately isolated to the RF amplifier. All other circuits are eliminated, but this is a lucky accident, not good troubleshooting. To be performed efficiently and rapidly, the troubleshooting procedures should be based on a systematic, logical process, not on chance or luck. You will probably have as many unlucky accidents as lucky ones throughout your troubleshooting career.

The same condition holds true if test point D is chosen first (if you use signal tracing). A normal signal at D clears all circuits except the noise limiter. An abnormal signal still leaves the possibility of trouble in many circuits.

If test point B is chosen first and the signal is normal, this clears three circuits (RF amplifier, mixer, local oscillator), leaves four circuits possibly defective (both IF amplifiers, detector, and noise limiter). You get the opposite results if the signal is abnormal at B. However, test point B is not a bad choice for a first test using signal tracing. If test point B is more readily accessible than test point C, use B.

Half-split technique using signal injection.   If signal injection is used, signals of the right sort (proper frequency and amplitude) are injected at test points A, B, C, D, and the antenna, as shown in Fig. 1–15. Receiver circuit response is noted at the volume control or at the loudspeaker. In this case, the signal for test points A and the antenna is at the RF frequency of the selected channel, modulated by an AF tone. The signals for B and C are at the IF frequency, also modulated by an AF tone. The signal for D is an AF tone. Note that signal injection requires several different types of signal sources at different frequencies, whereas signal tracing requires only one signal at the input.

Using signal injection with the half-split technique, the first signal is again injected at test point C. Now, however, a normal response at this volume control or loudspeaker clears the final circuits (detector and noise limiter). Under these conditions, the next logical points for signal injection are B and A, in that order.

An absent or abnormal response at the volume control or loudspeaker (with a signal injected at C) isolates the trouble to the detector or noise limiter. With such results, the next logical test point is D.

A normal response at the volume control or loudspeaker with a signal at D, but not with a signal at C, isolates the trouble to the detector. An absent or abnormal response with a signal at D isolates the trouble to the noise limiter.

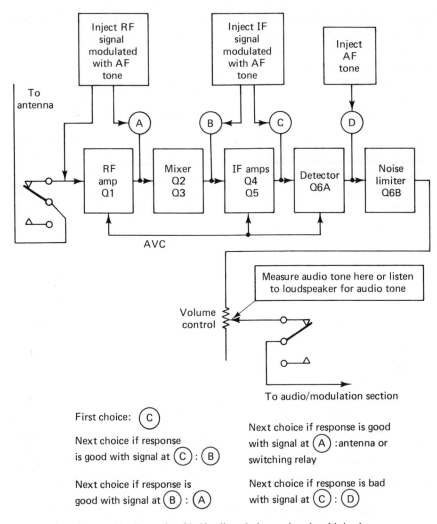

Figure 1-15   Example of half-split technique using signal injection.

Keep in mind that these examples using the half-split technique, signal tracing, signal injection, and bracketing to isolate the trouble to a circuit group by no means cover all the possibilities that may occur. They simply illustrate the basic concepts involved when following the systematic, logical troubleshooting procedures found throughout the author's books.

1-5.6 Isolating Trouble to a Circuit Within a Circuit Group

Once trouble is definitely isolated to a faulty circuit group, the next step is to isolate the trouble to the faulty circuit within the group. Bracketing, half-splitting, signal tracing, signal injection, and a knowledge of the signal path in the circuit group are all important in this step, and are essentially the same methods used for isolating trouble to the circuit group.

1-6. LOCATING A SPECIFIC PROBLEM

The ability to recognize symptoms and to verify them with test equipment helps you to make logical decisions regarding the selection and localization of the faulty circuit group. This ability also helps you to isolate trouble to a faulty circuit. The final step of troubleshooting—locating the specific trouble—requires testing of the various branches of the faulty circuit to find the defective part.

The proper performance of the locate step enables you to find the cause of trouble, repair it, and return the equipment to normal operation. A follow-up to this step is to record the trouble so that, from the history of the equipment, future troubles may be easier to locate. Also, such a history may point out consistent failures which could be caused by a design error.

1-6.1 Locating Trouble in Plug-in Modules

Because the trend in modern electronic equipment is toward IC and sealed-module design, technicians often assume that it is not necessary to locate specific troubles to individual parts. That is, they assume that all troubles can be repaired by replacement of sealed modules. Some technicians are even trained that way. The assumption is not always true.

While the use of replaceable modules often minimizes the number of steps required in troubleshooting, it is still necessary to check circuit branches to parts outside the module. Front-panel operating controls are a good example of this, since such controls are not located in the sealed units. Instead, the controls are connected to the terminals of an IC, circuit board, or plug-in module.

1-6.2 Inspection Using the Senses

After the trouble is isolated to a circuit, the first step in locating the trouble is to perform a preliminary inspection using the senses. For example, burned or charred resistors can often be spotted by sight or smell. The same holds true for oil-filled or wax-filled parts, such as some capacitors, coils, and transformers.

Overheated parts, such as hot transistor cases, can be located quickly by touch. The sense of hearing can be used to listen for high-voltage arcing between wires or wires and the chassis, for "cooking" or overloaded or overheated transformers, or for hum or lack of hum, whichever the case may be. Although all the senses are used, the procedure is referred to most frequently as a visual inspection.

1-6.3 Testing to Locate a Faulty Part

Testing vacuum tubes.   Vacuum tubes are relatively easy to replace (compared to transistors or ICs). For that reason, two common practices were developed in the early days of electronics. One practice is to test all vacuum tubes, by substitution, as a first step in troubleshooting. The other practice is to remove and test all tubes on a tube tester, as a first troubleshooting step. Neither practice is recommended.

With tube substitution, the usual procedure is to replace tubes one at a time until the equipment again works normally, then the last tube replaced is discarded, and all the other original tubes are reinserted in their respective sockets. There are several problems with this approach.

Some oscillator or high-frequency circuits may operate with one new tube and not with another because of the differences in interelectrode capacitance between the tube elements (a good tube may react like a bad tube). When removing or inserting the tubes, rocking or rotating them may result in bent pins or broken weld wires where the pins enter the envelope. If there is more than one bad tube in the equipment, substituting good tubes one at a time and reinserting the original tube before substituting the next tube does not locate the defective tube. Finally, if the replacement tube becomes defective immediately after substitution, there definitely is circuit trouble, and further trouble-shooting is required anyway.

Testing all tubes on a tube tester, as the first troubleshooting step, is also not recommended. Because this procedure has been followed so religiously in the past, the practice has led to the misconception that defective tubes are the cause of all or most equipment failures. Even if defective tubes cause 50% (or more) of all equipment failures, the process of removing tubes, checking them on a sometimes marginal tube tester, and replacing them with new tubes, as a first step without further circuit checking, is a waste of time (and is poor troubleshooting practice).

This does not mean that the tube should never be checked first, after the trouble is isolated to a circuit. For example, the power can be turned on and the tube filaments checked first for proper warm-up. If the tube envelopes are glass, a visual inspection shows whether the filament is burned out. For metal-envelope tubes, you can feel the envelope to find out whether the filament is lit.

This type of test may speed the troubleshooting effort by quickly locating a tube having a burned-out filament. If the tube does not warm up properly, remove the tube and check it on a tube tester or substitute a new (known-good) tube, whichever is most convenient. In either case, the complete circuit should be checked to determine if the tube burned out naturally from long use or from some trouble in the circuit. Simply replacing the tube without checking the rest of the circuit does not complete the location of trouble. You still must verify whether the burned-out tube is the cause or the effect of trouble.

The procedure just described works well for checking tubes when the tube filaments are connected in parallel. With parallel filaments, when the filament of one tube burns out, that tube (and only that tube) shows a bad (unlit) filament. The filaments of tubes connected in series present more of a problem. When one filament burns out, all the filaments in the spring are unlit. This condition is illustrated in Fig. 1-16 and is typical for many vacuum-tube circuits (particularly the older, tube-type radio receivers and television sets). With such circuits, it is more difficult to determine which filament in the string is the bad one.

Removing the tubes one at a time and checking their filaments for continuity with an ohmmeter are time consuming, and if care is not taken, the tube may be damaged during the test since the current from the ohmmeter (set on its lowest scale) may be high enough to burn out the filament.

A better test is to measure the voltage across the filament terminals of the tube socket (provided that the bottom of the socket is accessible and all the tubes are left in their sockets). All good filaments in the string will show zero voltage, but the one that is defective (burned out) shows the full voltage applied to the string, as shown in Fig. 1-16.

Testing solid-state and IC equipment.   Unlike vacuum tubes, most transistors, ICs, and solid-state diodes are not easily replaced. Thus, the old electronic troubleshooting procedure of replacing tubes at the first sign of trouble has not carried over into present-day solid-state equipment. Instead, solid-state circuits are analyzed by testing to locate faulty parts.

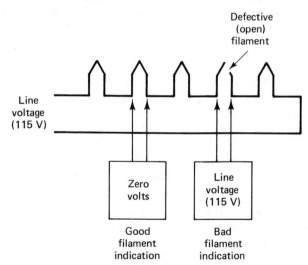

Figure 1-16   Method of locating defective (open) filaments in vacuum-tube equipment.

Testing the active device.   For troubleshooting purposes, the vacuum tube, transistor, IC, and solid-state diode may be considered the active device in any electronic equipment. Because of their key position in the circuit, these devices are a convenient point for evaluating operation of the entire circuit (through waveform, voltage, and resistance tests). Making these tests at the terminals of the active device usually results in locating the trouble quickly.

Waveform testing.   Usually, the first step in circuit testing is to analyze the output waveform of the circuit or the output of the active device (typically the collector, or possibly the emitter, or a transistor). Of course, in some circuits (such as power supplies) there is no output waveform (only a d-c voltage). And in some circuits there is no waveform of any significance.

In addition to checking for the presence of waveforms on an oscilloscope, the waveform must be analyzed in detail to check the amplitude, duration, phase, and/or shape. As discussed throughout this book, a careful analysis of waveforms can often pinpoint the most likely branch of a circuit that is defective.

Transistor and diode testers.   It is possible to test transistors and diodes in circuit, using in-circuit testers. These testers are usually quite good for transistors used at lower frequencies, particularly in the audio range. However, most in-circuit testers do not show the high-frequency characteristics of transistors. (The same is true for out-of-circuit transistor and diode testers.) For example, it is quite possible for a transistor to perform well in the audio range, or even in sweep circuits. However, the same transistor might be inadequate for RF, IF, or video circuits.

Voltage testing.   After waveform analysis and/or in-circuit tests, the next logical step is voltage measurement at the active device terminals or leads. Always pay particular attention to those terminals which show an abnormal waveform. These are the terminals most likely to show an abnormal voltage.

When properly prepared service literature is available (with waveform, voltage, and resistance information), the actual voltage measurements can be compared with the normal voltages listed in the service literature. This test often helps isolate the trouble to a single branch of the circuit.

Relative voltages.   It is often necessary to troubleshoot solid-state circuits without benefit of adequate voltage and resistance information. This can be done using the schematic diagram to make a logical analysis of the relative voltages at the transistor terminals. For example, with an NPN transistor, the base must be positive in relation to the emitter if there is to be emitter–collector current flow. That is, the emitter–base junction is forward-biased when the base is more positive (or less negative) than the emitter.

Resistance measurements.   After waveform and voltage measurements are made, it is often helpful to make resistance measurements at the same point on the active device (or at other points in the circuit), particularly where an abnormal waveform and/or voltage is found. Suspected parts often can be checked by a resistance measurement, or a continuity check can be made to find point-to-point resistance of the suspected branch. Considerable care must be used when making resistance measurements in solid-state circuits. The junctions of transistors act like diodes. When biased with the right polarity (by the ohmmeter battery), the diodes conduct and produce false resistance readings.

Current measurements.   In rare cases, the current in a particular circuit branch can be measured directly with an ammeter. However, it is usually simpler and more practical to measure the voltage and resistance of a circuit and then calculate the current.

1-6.4 Waveform and Signal Measurement

When testing to locate trouble, the waveform measurements are made with the circuit in operation and usually with an input signal (or signals) applied. The signals can be from an external generator, or you can use alternative sources (such as television broadcast signals or signals from another communications set). If you use the waveform reproductions found in the service literature, follow all the notes and precautions described in the literature. Usually, the literature will specify the position of operating controls, typical input signal amplitudes, and so on.

Note that most television circuit waveforms are measured with the oscilloscope sweep setting at 30 Hz or 7875 Hz. The 30-Hz frequency is one-half the vertical frequency of 60 Hz. There, there are two waveforms displayed on the oscilloscope when checking any circuit containing vertical pulses or signals (output of video detector, sync separator, vertical sweep). Similarly, the 7875-Hz frequency is one-half the horizontal frequency of 15,750 Hz and is used to display two waveforms when checking circuits containing horizontal signals (output of video detector, sync separator, horizontal oscillator, horizontal output).

Figure 1-17 shows waveform reproductions found in typical television service literature. That is, the approximate shape of the waveform is given on the schematic diagram (rather than in a separate chart), together with the voltage amplitude. There may or may not be a note on the diagram indicating the approximate frequency of the waveform. Note that only the horizontal sync pulses (15,750 Hz) are shown at the input to the sync amplifier Q1 (at the base). As a practical matter, it is reasonable to assume that there are vertical sync pulses at the same point. Thus, you could monitor the Q1 base with an oscilloscope and expect to see two waveforms representing the vertical sync pulses (if the oscilloscope is set to a 30-Hz sweep).

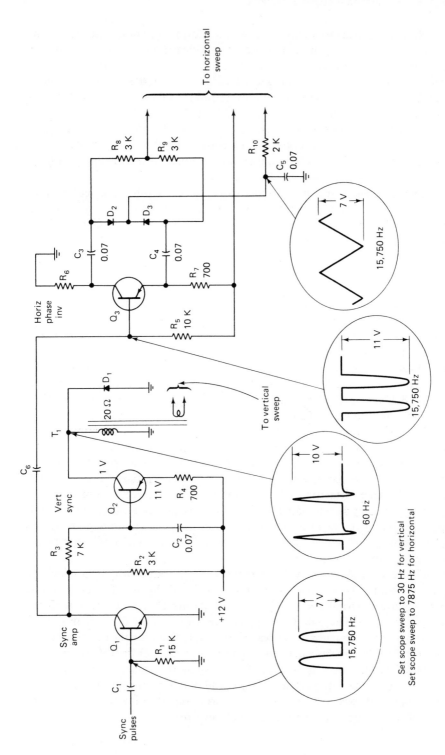

Figure 1-17 Waveform representations found on typical television service schematics.

There is a relationship between waveforms and trouble symptoms. Complete failure of a circuit usually results in the absence of a waveform. A poorly performing circuit usually produces an abnormal or distorted waveform. Also, exact waveforms are not always critical in all circuits. The representations of waveforms given in television service literature (and much other service literature) are usually only approximations of the actual waveform. Also, the same waveform does not always appear exactly the same when measured with different oscilloscopes.

1-6.5 Voltage Measurements

When testing to locate trouble the voltage measurements are made with the circuit in operation but usually with no signals applied. If you are using the voltage information found in service literature, follow all the notes and precautions. Usually, the information will specify the position of operating controls, typical input voltages, and so on.

In some electronic service literature, the voltages are given on the schematic diagram, together with the waveforms, as shown in Fig. 1-18. In other service literature, the voltage information is presented in chart form as shown on Fig. 1-19. Either system is quite accurate, but both systems require that you find the actual physical location of the terminals where the voltages are to be measured. (In some military-style service literature, voltage and resistance charts are arranged to simulate actual physical relationship or position of the terminals.)

Because of the safety practice of setting a voltmeter to its highest scale before making measurements, the terminals having the highest voltage should be checked first. (The order in which you check the voltages makes little difference when using an auto-ranging voltmeter. However, it is good practice to check the highest voltages first, to establish the habit.) Then the elements having less voltage should be checked in descending order.

If you have had any practical experience in troubleshooting, you know that voltage (as well as resistance and waveform) measurements are seldom identical to those listed in the service literature. This brings up an important question concerning voltage measurements: "How close is good enough?" In answering this question, there are several factors to consider.

The tolerances of the resistors, which greatly affect the voltage readings in a circuit, may be 20, 10, or 5%. Resistors with 1% (or better) tolerance are used in some critical circuits. The tolerances marked or color-coded on the parts are therefore one important factor. Transistors and diodes have a fairly wide range of characteristics and thus cause variations in voltage readings. The accuracy of test instruments must also be considered. Most voltmeters have accuracies of a few percent (typically 5 to 10%). Precision laboratory meters have a much greater accuracy.

For proper operation, critical circuits may require voltage to be within a

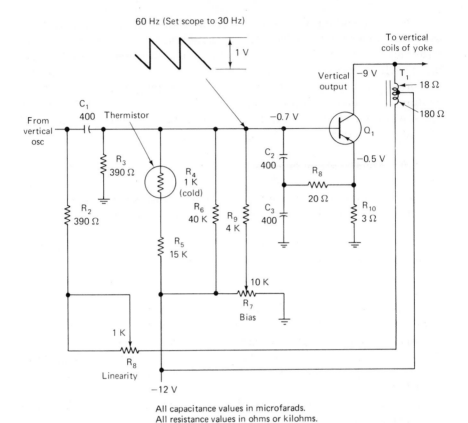

Figure 1-18 Voltage (and waveform) representations found on typical television service schematics.

SQUELCH AMPLIFIER DC VOLTAGE CHART

Transistor	Mode	DC Voltage In Volts		
		Emitter	Base	Collector
Q_8	squelch	10.0	8.8	9.5
	unsquelch	4.4	8.0	0
Q_9	squelch	1.6	2.2	1.6
	unsquelch	0	0	3.5

Figure 1-19 Example of voltage information found in CB service literature.

very close tolerance (at least within 10% and probably closer to 3%). However, many circuits operate satisfactorily if the voltages are to within 20 or 30%.

Generally, the most important factors to consider in voltage measurement accuracy are the symptoms and the output signal. If no output signal is produced, you should expect a fairly large variation of voltages in the trouble area. Trouble that results in circuit performance that is just out of tolerance may cause only a slight change in circuit voltages.

1-6.6 Resistance Measurements

Resistance measurements must be made with no power applied. However, in some cases, various operating controls must be in certain positions to produce resistance readings similar to those found in the service literature. This is particularly true of controls that have variable resistances.

Always observe any notes or precautions found on the service literature. In any circuit, always check that the filter capacitors are discharged before making resistance measurements. After all safety precautions and notes have been observed, measure the resistance from the terminals of the active device to the chassis (or ground) or between any two points that are connected by wiring or parts.

In some electronic service literature, resistance information is given on the schematic diagram, together with the waveforms and voltage information, as shown in Fig. 1-18. In other service literature, the resistance information is presented in chart form as shown in Fig. 1-20. Do not be surprised if you find service literature with little or no resistance information. Quite often, the only resistances given are the values of resistors and the d-c resistance of coils and transformers.

The reasoning for the omission of resistance values from various terminals (such as active device terminals) has some merit. If there is a condition in any active-device terminal circuit that produces an abnormal resistance (say, an open or shorted resistor, or a resistor that has changed drastically in value), the

SQUELCH AMPLIFIER RESISTANCE CHART

Transistor	Mode	Resistance In Ohms		
		Emitter	Base	Collector
Q_8	squelch	3.9 K	38 K	10.9 K
	unsquelch	500	38 K	0
Q_9	squelch	1 K	2.7 K	10.9 K
	unsquelch	1 K	2.7 K	10.9 K

Figure 1-20   Example of resistance information found in CB service literature.

voltage at that terminal will be abnormal. If such an abnormal voltage reading is found, it is then necessary to check out each resistance in the terminal circuit on an individual basis.

Because of the shunting effect of other parts connected in parallel, the resistance of an individual part or circuit may be difficult to check. In such cases, it is necessary to disconnect one terminal of the part being tested from the rest of the circuit. This leaves the part open at one end, and the value of resistance measured is that of the part only.

Keep in mind that when making resistance checks a zero reading indicates a short circuit, and an infinite reading indicates an open circuit. Also remember the effect of the transistor junctions (acting as a forward-biased diode when biased on).

1-6.7 Duplicating Waveform, Voltage, and Resistance Measurements

If you are responsible for service of one type or model of electronic equipment, it is strongly recommended that you duplicate all the waveform, voltage, and resistance measurements found in the service literature with your own test equipment. This should be done with known-good equipment that is operating properly. Then when you make measurements during troubleshooting you can spot even slight variations. Always make the initial measurements with test equipment that is normally used during troubleshooting. If more than one set of test equipment is used, make the initial measurements with all available test equipment, and record the variations.

1-6.8 Using Schematic Diagrams

Regardless of the type of trouble symptom, the actual fault can be traced eventually to one or more of the circuit parts (transistors, ICs, diodes, capacitors, etc.). The waveform–voltage–resistance checks then indicated which branch within a circuit is at fault. You must locate the particular part that is causing the trouble in the branch.

This requires that you be able to read a schematic diagram. These diagrams show what is inside the blocks on a block diagram and provide the final picture of equipment operation. Usually, you must troubleshoot electronic equipment with nothing but a schematic diagram. If you are fortunate, the diagram gives some voltages and waveforms.

Examples of using schematic diagrams in troubleshooting.   The following shows how a schematic diagram is used to locate a fault within a circuit. Although this example involves only selected circuits of a television receiver, the same basic troubleshooting principles apply to most nondigital electronic circuits.

Assume that the circuit of Fig. 1-17 is being serviced. The reasoning that led to this particular circuit is as follows. The symptom is a complete lack of vertical sync. The television picture rolls vertically and cannot be controlled by the vertical hold. All other functions, including the horizontal sync, are normal. This localizes the trouble to either the vertical sweep circuits or to the sync separator circuits. You isolate the trouble to the sync separator circuits by means of waveform measurement.

Your first waveform measurement is at the collector of Q2, which is (simultaneously) the vertical output of the sync separator and the input to the vertical sweep circuits. The waveform is absent or abnormal, and you place a bad-output bracket at the collector of Q2. All the remaining waveforms shown in Fig. 1-17 are normal. Thus you can assume that Q1 is functioning normally (Q1 is passing the horizontal sync pulses to Q3) and should also pass the vertical sync pulses. However, it is better not to assume anything. Instead, confirm that there are vertical sync pulses at both the input (base) and output (collector) of Q1. Note that neither of these pulses is shown on Fig. 1-17 (which is quite typical for the service literature of much electronic equipment).

To overcome this lack of information, set the oscilloscope sweep to 30 Hz, and measure whatever waveforms appear at the base and collector of Q1. If you find two pulses, they are the vertical sync pulses. The amplitude of these pulses should be approximately equal to the horizontal pulse amplitude. The base pulse should be about 7 V, with a 10-V pulse at the collector. Assume that this is the case in making your measurement. Now, it is reasonable to assume that Q1 is *probably* functioning normally and that you have a good input to Q2.

With the trouble isolated to Q2, you must now locate the specific fault in the Q2 circuit, using voltage and resistance measurements. The parts involved are R3, R4, C2, Q2, D1, and T1. Note that only two voltages are given for the terminals of the active device, Q2. These are 1 V for the collector and 11 V for the emitter. There is only a 1-V drop across the winding of T1 (and across D1), yet the normal waveform shows a pulse of about 10 V in amplitude. This indicates that Q2 is normally biased to (or beyond) cutoff and is switched on by pulses from the junction of R3 and C2 (which form the integrator).

To bias a PNP transistor to cutoff, the base must be less negative (or more positive) than the emitter. Since a positive 12-V supply is used, the base of Q2 must be more positive than the emitter (11 V) but less than 12 V. If you measure some voltage in the range 11 to 12 V from the base of Q2 to ground, it is probably correct, and R3 is probably good. (You can assume that R2 is good since it feeds Q1, and Q1 is normal.)

To clear the vertical integrator (R3 and C2) of any suspicion, measure the waveforms at the collector of Q1 and the base of Q2, using oscilloscope sweeps of 30 Hz and 7875 Hz. R3 and C2 function as a low-pass filter to prevent the horizontal sync pulses from passing to Q2. If R3 and C2 are doing their job, waveforms should appear at the base of Q2 only when the 30-Hz sweep is used. If no waveforms appear, or if both the horizontal and vertical waveforms are found at the base of Q2, then R3 and C2 are suspect.

Now assume that there is a good pulse at the base of Q2 and that you have thus narrowed the problem down to Q2, R4, D1, or T1. At this point, you could check Q2 by substitution, or with an in-circuit transistor tester, if convenient. However, you will probably do better to make voltage measurements at all terminals of Q2. This pinpoints any obvious part failures. For example, if R4 or T1 is open, all the voltages will be abnormal. If R4 is shorted, the emitter will be at the supply voltage (12 V) instead of at 11 V.

Keep in mind that voltage measurements alone may not solve the problem. For example, if D1 is shorted, or leaking badly, it is still possible to get a nearly-normal voltage reading at the Q2 collector but a poor pulse waveform. Also, voltage measurements in pulse circuits are sometimes ambiguous. The d-c voltage is given as 1 V, yet the pulse (or ac) voltage is 10 V. If the meter is set to measure dc, the pulses may increase the average d-c voltage (in some meters) so that an abnormal reading appears, even though the circuit is functioning normally. If D1 is open, the d-c voltage could be normal, but the pulse waveform will be abnormal. Diode D1 functions to remove any negative-going pulses at the collector of Q2. As another example, assume that the primary winding of T1 has partially shorted turns. This will produce a nearly normal d-c voltage but would drastically reduce the pulse output to the vertical sweep circuits.

Unless you have pinpointed the problem with waveform and voltage measurements, your last step is to make resistance measurements. As shown, no resistance values are given for any of the Q2 terminals. Thus, there is no point in measuring from the terminals to ground (as is standard practice in troubleshooting military-type equipment). Instead, you must check the resistance of each part on an individual basis.

To get accurate readings of individual parts, you must disconnect one lead from the remainder of the circuit. If not, the effect of solid-state devices in the circuit can further confuse the troubleshooting process. For example, assume that you are measuring the T1 primary winding resistance by connecting an ohmmeter across the winding. If the ohmmeter leads are connected so that the positive terminal of the ohmmeter battery is connected to the D1 anode (ground end), D1 will be forward biased, and the ohmmeter will read the combined D1 and T1 resistances. This problem can be eliminated by reversing the ohmmeter leads and measuring the resistance both ways. If there is a difference in the resistance values with the leads reversed, check the schematic for possible forward-bias conditions in diodes and transistor junctions in the associated circuit. Whenever practical, simply disconnect one lead of the part being measured.

1-6.9 Internal Adjustments During Troubleshooting

Adjustment of controls (both internal adjustment controls and operating controls) can affect circuit conditions. This may lead to false conclusions during troubleshooting.

For example, the bias on the base of vertical output transistor Q1 (in Fig. 1-18) is set by vertical bias potentiometer R7. In turn, the value of the bias determines the portion of the sweep used by Q1. Thus, the sweep voltages applied to the vertical deflection yoke are set, in part, by the vertical bias control. Adjustment of the vertical bias control affects both height and linearity of the vertical sweep. However, the main purpose of the vertical bias control is to compensate when a major part in the circuit (such as the output transistor, output transformer, or vertical deflection yoke) is replaced. The vertical bias control usually does not require adjustment during troubleshooting. On the other hand, the vertical linearity control R8 has a considerable effect on the linearity of the vertical sweep and often requires adjustment during the troubleshooting process.

These two extremes often lead some technicians to one of two unwise courses of action. First, the technician may launch into a complete alignment procedure (or whatever internal adjustments are available) once the trouble is isolated to a circuit. No internal control, no matter how inaccessible, is left untouched. The technician reasons that it is easier to make adjustments than to replace parts. While such a procedure eliminates improper adjustment as a possible fault, the procedure can also create more trouble than is repaired. Indiscriminate internal adjustment is the technician's version of "operator trouble."

At the other extreme, a technician may replace part after part, where a simple screwdriver adjustment will repair the problem. This usually means that the technician simply does not know how to perform the adjustment procedure or does not know what the control does in the circuit. Either way, a study of the service literature should resolve the situation.

To take the middle ground, do not make any internal adjustments during the troubleshooting procedure until trouble has been isolated to a circuit, and then only when the trouble symptom or test results indicate possible maladjustment.

For example, assume that the vertical oscillator is provided with a back-panel adjustment control (vertical hold) that sets the frequency of oscillation. If waveform measurements at the circuit show that the vertical oscillator is off frequency (not at 60 Hz), it is logical to adjust the vertical hold control. However, if waveform measurements show only a very low output (but on frequency), adjustment of the vertical hold control during troubleshooting could be confusing (and could cause further problems).

An exception to this rule is when the service literature recommends alignment or adjustment as part of the troubleshooting procedure. Generally, alignment or adjustment is checked after test and repair have been performed. This assures that the repair (parts replacement) procedure has not upset circuit adjustment.

1-6.10 Trouble Resulting from More Than One Fault

A review of all the symptoms and test information obtained thus far helps you verify the part located as the sole trouble or to isolate the faulty parts. This is true whether the malfunction of these parts is due to the isolated part or some entirely unrelated cause.

If the isolated bad part can produce all the normal and abnormal symptoms and indications that you have accumulated, you can logically assume that it is the sole cause of the trouble. If not, you must use your knowledge of electronics and of the circuit to find out what other part (or parts) could have become defective and produce all the symptoms.

When one part fails it often results in abnormal voltages or currents that could damage other parts. Trouble is often isolated to a faulty part, which is a result of an original trouble, rather than the source of trouble.

For example, assume that the troubleshooting procedure thus far has isolated a transistor as the cause of trouble. The transistor has been burned out by excessive current. The problem is a matter of finding how the excessive current was produced. There are several possibilities.

Excessive current in a transistor can be caused by an extremely large input signal, which will overdrive the transistor. This indicates a fault somewhere in the circuitry ahead of the input connection. Power surges (intermittent excessive outputs) from the power supply can also cause the transistor to burn out. In fact, power-supply surges are a common cause of transistor burnout. All of these conditions should be checked before placing a new transistor in the circuit. Some other typical malfunctions, together with their common causes, include:

1. Burned-out transistors caused by *thermal runaway*. An increase in transistor current heats the transistor, causing a further increase in current, resulting in more heat. This continues until the heat dissipation capabilities of the transistor are exceeded. Bias-stabilization circuits are generally included in most well-designed transistors to prevent thermal runaway.
2. Power-supply overload caused by a short circuit in some portion of the voltage distribution network.
3. Burned-out transistor caused by a shorted blocking capacitor.

4. Blow fuses caused by power-supply surges or shorts in filtering (power network).

It is obviously impractical to list all the common faults and their related causes that you may find in troubleshooting electronic equipment. Generally, when a part fails the cause is a circuit condition which exceeded the maximum ratings of the part. However, it is quite possible for a part to simply "go bad."

The circuit condition that causes a failure can be temporary and accidental, or it can even be a basic design problem (as indicated by a history of repeated failures). No matter what the cause, your job is to find the trouble, verify the source or cause, and then repair the trouble.

1-6.11 Repairing Troubles

In a strict sense, repairing the trouble is not part of the troubleshooting procedure. However, repair is an important part of the total effort involved in getting the equipment back into operation. Repairs must be made before the equipment can be checked out and made ready for operation.

Never replace a part if it fails a second time without making sure that the cause of trouble is eliminated. Preferably, the cause of trouble should be pinpointed before replacing a part the first time. However, this is not always practical. For example, if a resistor burns out because of an intermittent short, and you have cleared the short, the next step is to replace the resistor. However, the short could happen again, burning out the replacement resistor. If so, you must recheck every element and lead in the circuit.

When replacing a defective part, an exact replacement should be used if it is available. If not available, and if the original part is beyond repair, an equivalent or better part should be used. Never install a replacement part having characteristics or ratings inferior to those of the original.

Another factor to consider when repairing the trouble is that, if at all possible, the replacement part should be installed in the same physical location as the original, with the same lead lengths, and so on. This precaution is generally optional in most low-frequency or d-c circuits but must be followed for high-frequency applications. In RF, IF, and video circuits, changing the location of parts may cause the circuit to become detuned or otherwise out of alignment.

1-6.12 Operational Checkout

Even after the trouble is found and the faulty part is located and replaced, the troubleshooting effort is not necessarily complete. You should make an operational check to verify that the equipment is free of all faults and is performing properly again. Never assume that simply because a defective part is located and replaced, the equipment will automatically operate normally again. In practical troubleshooting, never assume anything; prove it. Operate the equipment through all of its modes. This ensures that one fault has not caused another. If practical, after you have checked the equipment, have the operator or customer check all operating modes.

When the operational check is completed, and the equipment is "certified" (by you and the customer) to be operating normally, make a brief record of the symptoms, faulty parts, and remedy. This is particularly helpful when you must troubleshoot similar equipment. Even a simple record of troubleshooting gives you a history of the equipment for future reference.

If the equipment does not perform properly during the operational checkout, you must continue troubleshooting. If the symptoms are entirely different, it may be necessary for you to repeat the entire troubleshooting procedure from the start. However, this is usually not necessary.

For example, assume that the equipment does not check out because a replacement transistor has detuned the circuit (say, an IF amplifier transistor has been replaced). When this is the case you can repair the trouble by IF alignment rather than returning to the first step of troubleshooting and repeating the entire procedure. Keep in mind that you have arrived at the defective circuit or component by a systematic procedure. Thus, retracing your steps, one at a time, is the logical method.

1-7. SAFETY PRECAUTIONS IN TROUBLESHOOTING

In addition to a routine operating procedure, certain precautions must be observed during operation of any electronic test equipment during service. Many of the precautions are the same for all types of test equipment; others are unique to special test instruments such as meters, oscilloscopes, and signal generators. Some of the precautions are designed to prevent damage to the test equipment or to the circuit where the service operation is being performed. Other precautions are to prevent injury to you. Where applicable, special safety precautions are included throughout the various chapters of this book. The following general safety precautions should be studied thoroughly and then compared to any specific precautions called for in the test equipment service literature and in the related chapters of this book.

1. Many service instruments are housed in metal cases. These cases are connected to the ground of the internal circuit. For proper operation, the ground terminal of the instrument should always be connected to the ground of the equipment being serviced. Make certain that the chassis of the equipment being serviced is not connected to either side of the a-c line or to any potential above ground. If there is any doubt, connect the equipment being serviced to the power line through an *isolation transformer*.

2. Remember that there is always danger in servicing equipment that

operates at hazardous voltages (such as the high voltages used in TV picture tubes and the CRTs of computer terminals). Remember this especially as you pull off the equipment back panel and apply power. Always make some effort to familiarize yourself with the equipment *before* troubleshooting, bearing in mind that high voltages may appear at unexpected points in a defective equipment.

3. It is good practice to remove power before connecting test leads to high-voltage points. (High-voltage probes are often provided with alligator clips.) It is preferable to make all service connections with the power removed. If this is impractical, be especially careful to avoid accidental contact with circuits and objects that are grounded. Working with one hand away from the equipment and standing on a properly insulated floor lessens the danger of electrical shock.

4. Filter capacitors may store a charge large enough to be hazardous. Discharge filter capacitors (after turning off the power) before attaching test leads.

5. Remember that leads with broken insulation offer the additional hazard of high voltages appearing at exposed points along the leads. Check test leads for frayed or broken insulation before working with them.

6. To lessen the danger of accidental shock, disconnect test leads immediately after the test is complete.

7. Remember that the risk of severe shock is only one of the possible hazards. Even a minor shock can place you in danger of more serious risks, such as a bad fall or contact with a source of higher voltage.

8. The experienced service technician guards continuously against injury and does not work on hazardous circuits unless another person is available to assist in case of accident.

9. Even if you have considerable experience with test equipment used in service, always study the service literature of any instrument with which you are not thoroughly familiar.

10. Even if you have had considerable experience with test equipment, always study the instruction manual of any instrument with which you are not familiar.

11. Never measure a voltage with a meter set to measure current or resistance. To do so will burn out the meter movement (on a typical VOM). Similarly, never measure a current with a meter set to measure resistance.

12. Always start voltage and current measurements on the *highest* voltage or current scale. Then switch to a lower range as necessary to obtain a good center-scale reading.

13. Do not attempt to measure a-c voltages or current with meters set to measure dc. This could damage the meter movement and will produce errors in the meter readings. No damage will result (usually, but consult the instruction manual) if d-c voltages or currents are measured with the meter set to measure ac. However, the readings will be in error.

14. Use only shielded probes. Never allow your fingers to slip down to the metal probe tip when the probe is in contact with a "hot" circuit.

15. Avoid operating test equipment in strong magnetic fields. Such fields can cause inaccuracy in meter movements, and can distort CRT displays. Most good-quality test instruments are well shielded against magnetic interference.

16. Most test instruments have some maximum input voltage and current specified in the instruction manual. Do not exceed this maximum. Also, do not exceed the maximum line voltage or use a different power frequency on those instruments which operate from line power.

17. Avoid vibration and mechanical shock. Most electronic test equipment is delicate.

18. Study the circuit under test before making any test connections. Try to match the capabilities of the test instrument to the circuit under test. For example, if the circuit under test has a range of measurements to be made (ac, dc, RF modulated signals, pulses, or complex waves), it may be necessary to use more than one instrument. Most meters will measure dc and low-frequency ac. If an unmodulated RF carrier is to be measured, use an RF probe. If a carrier to be measured is modulated with low-frequency signals, a demodulator probe must be used. If pulses, square waves, or complex waves (combinations of ac, dc, and pulses) are to be measured, a peak-to-peak reading meter, or an oscilloscope, will provide the only meaningful indications.

19. There are two standard *international operator warning symbols* found on some test instruments. One symbol, a *triangle with an exclamation point at the center,* advises the operator to refer to the operating manual before using a particular terminal or control. The other symbol, *zigzag line simulating a lightning bolt,* warns the operator that there may be dangerously high voltage at a particular location, or that there is a voltage limitation to be considered when using a terminal or control.

1-8. TEST EQUIPMENT IN TROUBLESHOOTING

Chapters 2 and 4 describe test equipment used in troubleshooting for communications and digital electronics, respectively. Because an understanding of test equipment is essential to efficient troubleshooting, it is strongly recommended that you study the corresponding test equipment chapter before reading the troubleshooting chapter (read Chapter 2 before you read Chapter 3, and so on).

In many cases, the test equipment used for troubleshooting one type of equipment is the same as for other fields of electronics. For example, most troubleshooting procedures are performed using meters, signal generators, frequency counters, oscilloscopes, power supplies, and assorted clips, patch-chords, and so on. Theoretically, all troubleshooting procedures can be performed using basic test equipment, provided that the oscilloscopes have the necessary gain and bandpass characteristics, the frequency counters cover the necessary range, the signal generators cover the appropriate frequencies, and so on.

However, there are many specialized test instruments that greatly simplify troubleshooting for the various fields of electronics. In the case of communications electronics, instruments such as the SWR meter, field strength meter, RF wattmeter, dummy load, spectrum analyzer, and FM deviation meter greatly simplify various phases of troubleshooting. For TV, the sweep, analyst, pattern, and color generators, as well as the vectorscope, make life much easier. In the case of digital troubleshooting, the logic probe, pulser and clip, current tracer, and logic-state analyzer are absolutely essential. We concentrate on such specialized equipment throughout the various test equipment chapters.

It is not the purpose of this book to promote one type of test and troubleshooting equipment over another (or one manufacturer over another). Instead, the troubleshooting chapters are devoted to the basic operating principles or features of test/troubleshooting instrument types in common use. When you have studied the information, you may then select the type of test equipment best suited to your own troubleshooting needs and pocketbook.

Although complicated theory has been avoided, the discussions in the test equipment chapters cover the way each type of instrument is used in troubleshooting, and what signals or characteristics are to be expected from each. We also describe how the features and outputs found on present-day test equipment relate to specific problems in troubleshooting.

A thorough study of the test equipment chapters will familiarize you with the basic principles and operating procedures for typical equipment used in advanced troubleshooting. It is assumed that you will take the time to become equally familiar with the principles and operating controls for any particular test/troubleshooting equipment that you use. Such information is contained in the user's manuals for the particular equipment. It is absolutely essential that you become thoroughly familiar with your own equipment. No amount of textbook instruction makes you an expert in operating test equipment; it takes actual practice.

It is strongly recommended that you establish a routine operating procedure, or sequence of operation, for each item of test/troubleshooting equipment. This approach saves time and familiarizes you with the capabilities and limitations of your own equipment, thus minimizing the possibility of false conclusions based on unknown operating conditions.

2

Communications
Test
Equipment

This chapter describes both general and special test equipment used for communications troubleshooting.

2-1. SIGNAL GENERATORS

The signal generator is an indispensable tool for advanced communications troubleshooting. Without any type of signal generator you are entirely dependent on signals transmitted by another communications set, and you are limited to signal tracing only. This means that you have no control over frequency, amplitude, or modulation of the signals and have no means of signal injection. With a signal generator of the appropriate type, you can duplicate transmitted signals or produce special signals required for alignment and test of all circuits found in communications equipment. Also, the frequency, amplitude, and modulation characteristics of the signals can be controlled so that you can check operation of the receiver circuits under various signal conditions (weak, strong, normal, or abnormal signals).

2-1.1 Signal Generator Basics

An oscillator (audio, RF, pulse, etc.) is the simplest form of signal generator. At the most elementary level of troubleshooting, a single-stage AF or RF oscillator can serve the purpose of providing a signal source. The special test sets described in Sec. 2-13 generally include such basic oscillator circuits.

Beyond the simplest troubleshooting, most comprehensive communications service requires an RF and AF generator, and possibly a pulse generator. Another instrument that may be useful in communications work is the probe-type (or pencil-type) generator.

2-1.2 Probe or Pencil Generators

These generators (also known as pencil-type noise generators, signal injectors, and by various other names) are essentially solid-state pulse generators or oscillators with a fast-rise waveform output and no adjustments. The fast-rise output produces simultaneous signals over a wide frequency range. The output signals may be used to troubleshoot the receiver and audio/modulation sections of communications equipment. However, except in basic troubleshooting situations, such an instrument has many obvious drawbacks.

For example, to check the selectivity of the receiver circuits, the signal source must be variable in amplitude. To check the detector or audio portions of the receiver circuits, the signal source must be capable of internal and/or external modulation. These characteristics are not available in the pencil-type unit. As a result, even the least expensive shop-type (or even kit-type) generators have many advantages over the pencil generators.

2-1.3 RF Signal Generators

There are no basic differences between shop-type and lab-type generators. That is, both instruments produce RF signal capable of being varied in frequency and amplitude, and capable of internal and external modulation. However, the lab-type instruments have several refinements not found in shop equipment, as well as a number of quality features (this accounts for the wide difference in price). Following is a summary of the differences between shop and lab RF generators.

Output meter.   In most shop generators, the amplitude of the RF output is either unknown or approximated by means of dial markings. The lab generator has an output meter. The meter is usually calibrated in microvolts so that the actual RF output may be read directly. If you use a shop generator without a built-in output meter, you must monitor the output with an external meter. Keep in mind that the meter must be capable of indicating output signals on the order of a few microvolts to perform properly all receiver checks of some communications sets. Meters are discussed in Sec. 2-3.

Percentage-of-modulation meter.   Most shop generators have a fixed percentage of modulation (usually about 30%). Lab generators provide for a variable percentage of modulation and a meter to indicate this percentage. Some generators have two meters (one for output amplitude and one for modulation percentage). Some generators use the same meter for both functions.

Output uniformity.   Shop generators vary in output amplitude from band to band. Also, shop generators usually cover part of their frequency range by means of harmonics. Lab generators have a more uniform output over their entire operating range and cover the range with pure fundamental signals.

Wideband modulation.   Often, the oscillator of a shop generator is modulated directly. This can result in undesired frequency modulation. The oscillator of a lab generator is never modulated directly (unless it is designed to produce an FM output); instead, the oscillator is fed to a wideband amplifier where the modulation is introduced. Thus, the oscillator is isolated from the modulating signal.

Frequency or tuning accuracy.   The accuracy of the frequency or tuning dials for a typical shop generator is about 2 or 3%, whereas a lab generator has from about 0.5 to 1% accuracy. However, neither instrument can be used as a frequency standard for troubleshooting communications equipment, since required FCC accuracy is much greater. For example, the FCC requires an accuracy of 0.005% or better (preferably 0.0025%) for CB equipment. There are laboratory instruments generally described as *communications monitors* or frequency meter/signal generators that provide signals with accuracies of up to 0.00005%. These instruments are designed for commercial communications work (radio and TV broadcast, etc.), are quite expensive, and are thus not usually found in a typical radio communications service shop.

Combined frequency counter/signal generator.   To overcome the accuracy problem in practical communications troubleshooting, the simplest approach is to monitor the signal generator output with a frequency counter. Such counters are discussed in Sec. 2-5. Using this technique, the frequency of the RF signal is determined by the accuracy of the counter, not the generator.

Frequency range.   Obviously, any RF generator used in communications troubleshooting must be capable of producing signals at all frequencies used in the communications set. It is also convenient if the generator can produce signals at both harmonic and subharmonic frequencies.

Frequency drift.   Because a signal generator must provide continuous tuning across a given range, some type of variable-frequency oscillator (VFO) must be used. As a result, the output is subject to drift, instability, modulation (by noise, mechanical shock, or power-supply ripple), and other problems associated with VFOs. Frequency instability does not present too great a problem in practical communications troubleshooting, provided that you monitor the signal generator output with a frequency counter. Of course, continuous drift can be annoying. For this reason, lab generators have temperature-compensated capacitors, frequency synthesizers, and PLL (phase-locked-loop) circuits to minimize drift. Similarly, the effects of line-voltage variations are offset by regulated power supplies.

Shielding.   The better generators have more elaborate shielding, especially for the output-attenuator circuits, where RF signals are most likely to leak. The leakage of RF from signal generators is something of a problem during receiver circuit sensitivity tests, or any tests involving low-amplitude (microvolt range) RF signals from the generator.

Band spread.   Shop generators usually have a minimum number of bands for a given frequency range. This makes the tuning-dial or frequency-control adjustments more critical, as well as difficult to see. Lab generators usually have a much greater band spread; that is, they cover a smaller part of the frequency range in each band.

2-1.4 Audio Generators

Audio generators (often called *audio oscillators* by old-timers in radio communications) are useful in troubleshooting the audio/modulation circuits of a communications set. Audio generators may also be used as modulation sources for RF signal generators. For example, if your particular RF generator has only a 400-Hz internal modulation provision (which is typical), and the test requires 1000 Hz (also typical), you can modulate the RF generator with an audio generator tuned to 1000 Hz.

As in the case of RF signal generators, audio generators in their simplest form are essentially audio oscillators. For troubleshooting purposes, the audio output is tunable in frequency over the entire audio range (and beyond) and is variable in amplitude.

Early audio generators produced only sine waves. However, most present-day audio generators also produce square waves at audio frequencies. Some lab audio generators are referred to as *function generators* because they produce various functions: sine waves, square waves, triangular waves, and/or sawtooth waves. Only the sine waves are of any particular value in communications work. However, almost any audio generator available today has some of the other outputs.

The major differences in audio generators are in quality rather than in special features. For example, the better audio generators are less subject to frequency drift and line-voltage variations. The effects of hum or other line noises are minimized by extensive filtering. Accuracy and dial resolution are generally better in lab generators. This makes the tuning-dial adjustments less critical. Lab generators also have a more uniform output over their entire operating range, whereas shop-type generators may vary in amplitude from band to band.

Keep in mind that if you want accuracy from an audio generator, you must monitor the output signal with a meter (for signal amplitude) and a frequency counter (for signal frequency).

2-1.5 Pulse Generators

The most common use for a pulse generator in communications trouble-shooting is the test and adjustment of *noise blanker* circuits. Some communications receivers have blanking circuits that detect noise signals or pulses at the receiver input (antenna), and function to desensitize (or cut off) the receiver in the presence of large noise signals. A pulse generator may be used to simulate noise bursts. However, the service literature for every communications set with noise blanking circuits does not always recommend a pulse generator. In many cases, the noise blanking circuits are tested and adjusted with an RF generator, modulated by an AF generator.

2-2. OSCILLOSCOPES

There are two uses for oscilloscopes in communications troubleshooting: signal tracing and modulation measurement.

2-2.1 Signal Tracing with an Oscilloscope

The signals in all circuits of a communications set may be traced with an oscilloscope, provided that the scope is equipped with the proper probe (probes are discussed further in Sec. 2-4). You can check amplitude, frequency, and waveforms of the signals with a scope. However, many communications service technicians do not use scopes extensively, for the following reasons.

The oscilloscope can measure signal amplitude, but a meter is easier to read. The same applies to signal frequency. The frequency counter is easier to read, and it is far simpler to measure frequency with a meter than with a scope, particularly in the typical communications frequency range. The oscilloscope is a superior instrument for monitoring waveforms. However, most communications signals are sine waves, and waveforms are not critical.

2-2.2 Modulation Checks with an Oscilloscope

The main use for an oscilloscope in communications work is to measure the percentage of modulation and uniformity or linearity of modulation. The use of an oscilloscope for modulation checks is not new. There are many variations of the basic technique, each of which is discussed in the following.

2-2.3 Direct Measurement of the Modulation Envelope with a High-Frequency Oscilloscope

If the vertical channel response of the oscilloscope is capable of handling the transmitter output frequency, the output can be applied directly through the oscilloscope vertical amplifier. The basic test connections are shown in Fig. 2-1. The procedure is as follows:

1. Connect the oscilloscope to the antenna jack, or the final RF amplifier of the transmitter, as shown in Fig. 2-1. Use one of the three alternatives shown, or the modulation measurement described in the transmitter service literature.

2. Key the transmitter (press the push-to-talk switch) and adjust the oscilloscope controls to produce displays as shown. You can either speak into the microphone (for a rough check of modulation), or you can introduce an audio signal (typically at 400 or 1000 Hz) at the microphone jack input (for a precise check of modulation). Note that Fig. 2-1 provides simulations of typical oscilloscope displays during modulation tests.

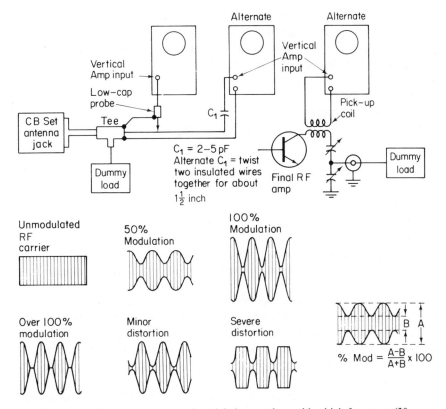

Figure 2-1   Direct measurement of modulation envelope with a high-frequency (30 MHz or higher) oscilloscope.

3. Measure the vertical dimensions shown as A and B in Fig. 2–1 (the crest amplitude and the trough amplitude). Calculate the percentage of modulation using the equation of Fig. 2–1. For example, if the crest amplitude (A) is 63 (63 screen divisions, 6.3 V, and so on) and the trough amplitude (B) is 27, the percentage of modulation is

$$\frac{63 - 27}{63 + 27} \times 100 = 40\%$$

Make certain to use the same oscilloscope for both crest (A) and trough (B) measurements. Keep in mind when making modulation measurements, or any measurement that involves the transmitter, that the RF output (antenna connector) must be connected to an antenna or a dummy load. Dummy loads are discussed in Sec. 2–6.

2–2.4 Direct Measurement of the Modulation Envelope with a Low-Frequency Oscilloscope

If the oscilloscope is not capable of passing the transmitter frequency, the transmitter output can be applied directly to the vertical deflection plates of the oscilloscope cathode-ray tube. However, there are two drawbacks to this approach. First, the vertical plates may not be readily accessible. Next, the voltage output of the final RF amplifier may not produce sufficient deflection of the oscilloscope trace.

The test connections and modulation patterns are essentially the same as those shown in Fig. 2–1. Similarly, the procedures are the same as those described in Sec. 2–2.3.

2–2.5 Trapezoidal Measurement of the Modulation Envelope

The trapezoidal technique has an advantage in that it is easier to measure straight-line dimensions than curving dimensions. Thus, any nonlinearity in modulation may easily be checked with a trapezoid. In the trapezoidal method, the modulated carrier amplitude is plotted as a function of modulating voltage, rather than as a function of time. The basic test connections are shown in Fig. 2–2.

1. Connect the oscilloscope to the final RF amplifier and modulator. As shown in Fig. 2–2, use either the capacitor connection or the pickup coil for the RF (oscilloscope vertical input). However, for best results, connect the transmitter outputs directly to the deflection plates of the oscilloscope tube. The oscilloscope amplifiers may be nonlinear and can cause the modulation to appear distorted.

2. Key the transmitter and adjust the controls (oscilloscope controls and R1) to produce a display as shown.

3. Measure the vertical dimensions shown as A (crest) and B (trough) on

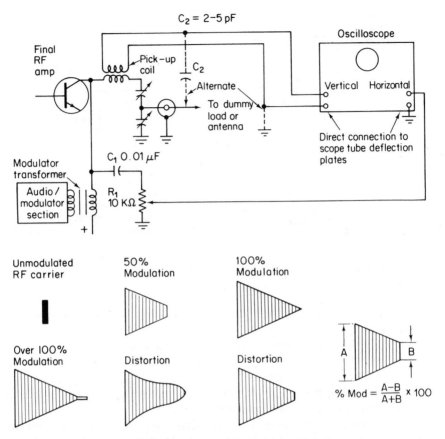

Figure 2-2 Trapezoidal measurement of the modulation envelope.

Fig. 2-2, and calculate the percentage of modulation using the equation given. For example, if the crest amplitude A is 80, and the trough amplitude B is 40, using the same scale, the percentage of modulation is

$$\frac{80 - 40}{80 + 40} \times 100 = 33\%$$

Again, make sure that the transmitter output is connected to an antenna or dummy load, before transmitting.

2-2.6 Down-Conversion Measurement of the Modulation Envelope

If the oscilloscope is not capable of passing the transmitter carrier signals, and the transmitter output is not sufficient to produce a good indication when connected directly to the oscilloscope tube, it is possible to use a down-converter test setup. One method requires an external RF generator and an IF

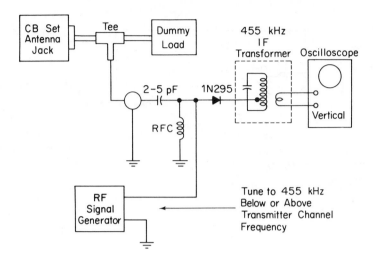

Figure 2–3   Down-conversion method of modulation measurement using a 455-kHz
IF transformer.

transformer. The other method uses a receiver capable of monitoring the
transmitter frequencies.

The RF generator method of down-conversion is shown in Fig. 2–3. In
this method, the RF generator is tuned to a frequency above or below the
transmitter frequency by an amount equal to the IF transformer frequency. For
example, if the IF transformer is 455 kHz, tune the RF generator to a frequency
455 kHz above (or below) the transmitter frequency.

The receiver method of down-conversion is shown in Fig. 2–4. With this
method, the receiver is tuned to the transmitter frequency, and the oscilloscope
input signal is taken from the last IF stage output through a 30-pF capacitor.

With either method of down-conversion, the RF generator or receiver is
tuned for a maximum indication on the oscilloscope screen. Once a good pattern is obtained, the rest of the procedure is the same as described in Sec. 2–2.3.

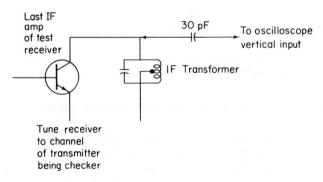

Figure 2–4   Down-conversion method of modulation measurement using a receiver.

The author does not generally recommend the down-conversion methods, except as a temporary measure. There are a number of relatively inexpensive oscilloscopes available that will pass signals up to and beyond the 50-MHz range.

2–2.7 Linear Detector Measurement of the Modulation Envelope

If you must use an oscilloscope that does not pass the carrier frequency of the transmitter, you can use a linear detector. However, the oscilloscope must have a d-c input, where the signal is fed directly to the oscilloscope vertical amplifier, not through a capacitor. Most modern oscilloscopes have both a-c (with capacitor) and d-c inputs. The basic test connections for linear detection of the modulation envelope are shown in Fig. 2–5. The basic test procedure is as follows:

1. Connect the transmitter output to the oscilloscope through the linear detector circuit as shown in Fig. 2–5. Make certain to include the dummy load (or wattmeter) as shown.

2. With the transmitter not keyed, adjust the oscilloscope position con-

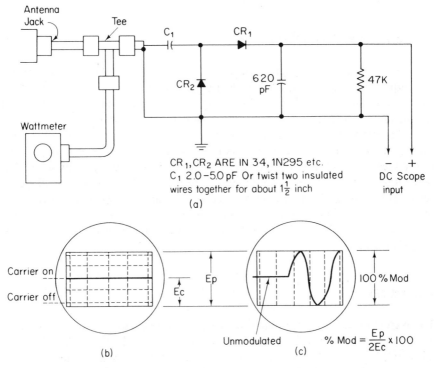

Figure 2–5   Pace modulation detector for measurement of the modulation envelope with a low-frequency oscilloscope.

trol to place the trace on a reference line near the bottom of the screen, as shown in Fig. 2–5(b) (carrier off).

3. Key the transmitter, but do not apply modulation. Adjust the oscilloscope gain control to place the top of the trace at the center of the screen, as shown in Fig. 2–5(b) (carrier on). It may be necessary to switch the transmitter off and on several times to adjust the trace properly, since the position and gain controls of most oscilloscopes interact.

4. Measure the distance (in scale divisions) of the shift between the carrier-on (step 3) and carrier-off (step 2) traces. For example, if the screen has a total of 10 vertical divisions, and the no-carrier trace is at the bottom or zero line, there is a shift of five scale divisions to the centerline.

5. Key the transmitter and apply modulation. Do not touch either the position or gain controls of the oscilloscope.

6. Find the percentage of modulation using the equation shown in Fig. 2–5. For example, assume that the shift between the carrier and no-carrier trace is five divisions, and that the modulation produces a peak-to-peak envelope of eight divisions. The percentage of modulation is

$$\frac{8}{2 \times 5} \times 100 = 80\%$$

2–2.8 Modulation Nomogram

Figure 2–6 is a nomogram that can be used with the direct measurement techniques (Secs. 2–2.3 and 2–2.4) or the trapezoidal technique (Sec. 2–2.5) to find percentage of modulation. To use Fig. 2–6, measure the values of the crest (or maximum) and trough (or minimum) oscilloscope patterns.

The percentage of modulation is found by extending a straightedge from the measured value of the crest or maximum (given as A on Fig. 2–6) on its scale to the measured value of the trough or minimum (given as B) on its scale. The percentage of modulation is found where the straightedge crosses the diagonal scale. The crest and trough may be measured in any units (volts, vertical scale divisions, etc.) as long as both crest and trough are measured in the same units. The dashed line in Fig. 2–6 is used to illustrate the percentage of modulation example of Sec. 2–2.3.

2–3. METERS

The meters used for communications troubleshooting are essentially the same as for all other electronic service fields. Most tests can be done with the standard VOM (volt-ohmmeter) or multimeter. The VOM may be either digital or moving-needle. Some of the digital meters with multiple functions require 115-V line power, and are thus best suited for use in the shop. Most moving-

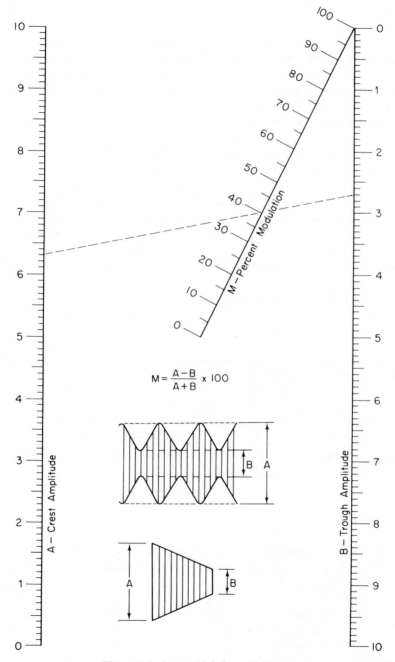

$$M = \frac{A - B}{A + B} \times 100$$

Figure 2-6 Pace modulation nomogram.

needle VOMs are difficult to read (in comparison to digital) but operate on internal batteries, and can thus be used in the shop or in the field. Some digital meters do not have dB (decibel) scales.

The meters can be used to measure both voltages and resistances of communications circuits, as required for the troubleshooting procedures described in Chapter 3. When used with the appropriate probe (Sec. 2–4), the meters can be used to trace signals throughout all communications circuits, including receiver (RF and IF), transmitter, and audio/modulator circuits. When used with the correct probe, the meter indicates the presence of a signal in the circuit and the signal amplitude, but not the signal frequency or waveform.

In addition to accuracy, ranges (both high and low), and resolution or readability, meters are rated in terms of ohms per volt; 20,000 ohms per volt is typical. A higher ohms-per-volt rating means that the meter draws less current and thus has the least disturbing effect on the circuit under test. A lower ohms-per-volt rating means more circuit loading, which should be avoided in some critical circuits. For example, the AVC or AGC circuits of some communications receivers will not operate properly when loaded with a low ohms-per-volt meter. The same is true of some oscillator circuits found in communications transmitters.

One way to avoid the loading problem is to use an electronic voltmeter that has a high input impedance and thus draws very little current from the circuit under test. The electronic voltmeter can be VTVM (vacuum-tube voltmeter), EVM (electronic voltmeter), TVM (transistorized voltmeter), or some similar instrument. Most digital meters are electronic meters and thus draw a minimum of current from the circuit.

One minor problem with some meters is that the frequency range is not sufficient to cover the entire audio range, which is usually considered anything up to about 20 kHz. A typical meter will have a 10-kHz maximum range, without the use of a probe. The problem may be overcome with a probe of the proper type. Also, the range of the audio/modulation circuits of a typical communications set is about 3 kHz maximum.

2–4. PROBES

In practical communications troubleshooting, all meters and oscilloscopes operate with some type of probe. In addition to providing for electrical contact to the circuit being tested, probes serve to modify the voltage being measured to a condition suitable for display on an oscilloscope or readout on a meter.

In its simplest form, the basic probe is a *test prod*. Sometimes, the probe tip is provided with an alligator clip so that it is not necessary to hold the probe at the circuit point. Basic probes work well on communications circuits carrying d-c and audio signals. However, if the alternating current being monitored is at a high frequency, or if the gain of the meter (such as an electronic meter) or scope amplifier is high, it may be necessary to use a special *low-capacitance probe.* Hand capacitance in a basic probe or test prod can cause hum pickup, particularly if amplifier gain is high. This condition may be offset by shielding in low-capacitance probes. More important, however, is the fact that the input impedance of the meter or scope is connected directly to the circuit being tested when a basic probe is used. Such impedance may disturb circuit conditions as discussed in Sec. 2-3.

High-voltage probes are rarely, if ever, used in communications troubleshooting. Most present-day communications sets are solid-state and operate with voltages of less than 15 V. Even the vacuum-tube sets generally use voltages of 300 V or less. Most VOMs will easily handle voltages in this range. However, some meters are supplied with high-voltage probes, either as accessories or built-in. If you should use these probes, the voltage indications will be 10:1, 100:1, or even 1000:1, depending on the attenuation factor. Also, note that some low-capacitance probes serve the dual purpose of capacitance reduction and voltage reduction. You must remember that voltage indications will be one-tenth (or whatever value of attenuation is used) of the actual value when such probes are used.

When the signals to be measured are at radio frequencies and are beyond the frequency capabilities of the meter or scope circuit, an *RF probe* is required. RF probes convert (rectify) the RF signals to a d-c output voltage that is equal to the peak RF voltage. The d-c output of the probe is then applied to the meter or scope input and is displayed as a voltage readout in the normal manner. In some RF probes, the d-c output is equivalent to peak RF voltage, whereas in other probes, the readout is equal to rms voltage. A few RF probes provide peak-to-peak values.

The circuit of a *demodulator probe* is essentially the same as that of the RF probe, but the circuit values and basic functions are somewhat different. The demodulator probe produces a-c and d-c outputs. The RF carrier frequency is converted to a d-c voltage equal to the peak of the RF carrier. The low-frequency modulating voltage (if any) appears as ac at the probe output.

In troubleshooting with a demodulator probe, the meter or scope is set to measure direct current and the RF carrier is measured. Then the meter or scope is set to measure alternating current, and the modulating voltage is measured. In general, demodulator probes are used primarily for signal tracing (as part of troubleshooting), and their output is not calibrated to any particular value.

It is possible to increase the sensitivity of a probe with a *transistor amplifier.* Such an arrangement is particularly useful with a basic VOM for measuring small-signal voltages during troubleshooting. An amplifier is usually not required for an electronic meter or scope because such instruments contain built-in amplifiers.

Probes must be calibrated to provide a proper output to the meter or scope with which they are to be used. Probe compensation and calibration are best done at the factory, using proper test equipment. Never attempt to adjust a probe unless you follow the instruction manual. An improperly adjusted probe will produce erroneous readings, and may cause undesired circuit loading.

2-4.1 Probe Troubleshooting Techniques

Although a probe is a simple instrument and does not require specific operating procedures, several points should be considered in order to use a probe effectively in troubleshooting.

Circuit loading.   When a probe is used, the probe's impedance (rather than the meter's or the scope's impedance) determines the amount of circuit loading. As discussed, connecting a meter or scope to a circuit may alter the signal at the point of connection. To prevent this, the impedance of the measuring device must be large in relation to that of the circuit being tested. Thus, a high-impedance probe offers less circuit loading, even though the meter or scope may have a lower impedance.

Measurement error.   The ratio of the two impedances (of the probe and the circuit being tested) represents the amount of probable error. For example, a ratio of 100:1 (perhaps a 100-MΩ probe to measure the voltage across a 1-MΩ circuit) produces an error of about 1%.

Effects of frequency.   The input impedance of a probe is not the same at all frequencies. (Capacitive reactance and impedance decrease with an increase in frequency.) All probes have some input capacitance. Even an increase at audio frequencies may produce a significant change in impedance.

Shielding capacitance.   When using a shielded cable with a probe to minimize pickup of stray signals and hum, the additional capacitance of the cable should be considered. The capacitance effects of a shielded cable can be minimized by terminating the cable at one end in its characteristic impedance. Unfortunately, this is not always possible with the input circuit of most meters and scopes.

Relationship of loading to attenuation factor.   The reduction of loading (either resistive or capacitive) due to the use of probes may not be the same as the attenuation factor of the probe. (Capacitive loading is almost never reduced by the same amount as the attenuation factor because of the additional capacitance of the probe cable.) For example, a typical 5:1 attenuator probe may be able to reduce capacitive loading by about 2:1. A 50:1 attenuator probe may reduce capacitive loading by about 10:1. Beyond this point, little improvement can be expected because of stray capacitance at the probe tip.

Checking effects of the probe.   When troubleshooting, it is possible to check the effect of a probe on a circuit by making the following simple test: Attach and detach another connection of a similar kind (such as another probe) and observe any difference in meter reading or scope display. If there is little or no change when the additional probe is touched to the circuit, it is safe to assume that the probe has little effect on the circuit.

Probe length and connections.   Long probes should be restricted to the measurement of relatively slow changing signals (direct current and low-frequency ac). The same is true for long ground leads. The ground leads should be connected where no hum or high-frequency signal components exist in the ground path between that point and the signal pick-off point.

Measuring high voltages.   Avoid applying more than the rated voltage to a probe. Fortunately, most commercial probes will handle the highest voltages found in communications circuits.

2-4.2 An RF Probe for Communications Troubleshooting

Figure 2-7 shows the schematic diagram (with circuit values) of a probe suitable for general communications troubleshooting. The probe, designed specifically for use with a VOM or electronic voltmeter, converts AF and RF signals to direct current. The RF signals can be at any frequency up to 250 MHz and possibly above. Note that the 47-kΩ resistor is not used with a VOM. Also, the probe is essentially a signal-tracing device and is not designed to provide accurate readings. The meter used with the probe must be set to read direct current, because the probe output is dc. However, if the RF input signal is modulated, the probe output may be pulsating direct current.

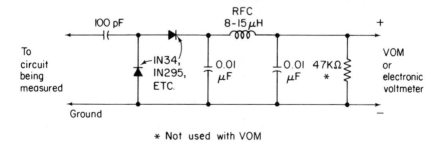

* Not used with VOM

Figure 2-7   Pace RF probe for CB service.

2-5. FREQUENCY METERS AND COUNTERS

There are two basic types of frequency-measuring devices for communications troubleshooting: the heterodyne or zero-beat frequency meter and the digital electronic counter.

2-5.1 Heterodyne or Zero-Beat Frequency Meter

In the early days of radio communications, the heterodyne meter was the only practical device for frequency measurement of transmitter signals. Figure 2–8 shows the block diagram of a basic heterodyne frequency meter. The signals to be measured are applied to a mixer, together with the signals of a known frequency (usually from a variable-frequency oscillator in the meter). The meter oscillator is adjusted until there is a null or "zero beat" on the output device, indicating that the oscillator is at the same frequency as the signals to be measured. This frequency is read from the oscillator frequency control dial. Precision frequency meters often include charts or graphs to help interpret frequency dial readings, so that exact frequencies can be pinpointed.

As an alternative system, the meter produces fixed frequency signals that are applied to the mixer, together with the signals of unknown frequency. As an example, one such frequency meter provides 40 crystal-controlled signals, one signal for each of the 40 CB channels. Both the CB transmitter and frequency meter are set to the same channel, and any deviation is read out on the frequency meter indicator.

2-5.2 Electronic Digital Counter

The electronic counter has become far more popular for communications work than the heterodyne frequency meter. One reason is that the counter is

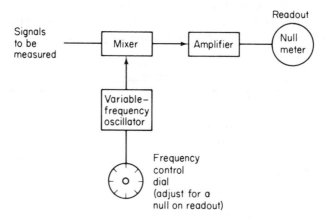

Figure 2-8   Basic heterodyne or zero-beat frequency meter circuit.

generally easier to operate and has much greater resolution or readability. Using the counter, you need only connect the test leads to the circuit or test point, select a time base and attenuator/multiplier range, and read the signal frequency on a convenient digital readout.

Digital counter basics.   Although there are many types of digital counters, all counters have several basic functional sections in common. These sections are interconnected in a variety of ways to perform the various counter functions. Figure 2-9 shows the basic counter circuit for *frequency measurement operation* (which is the most common of the various counter functions used in communications work). A typical digital counter will also provide *totalize operation* (where the instrument adds up events) and *period operation* (where the instrument measures intervals up to a given time). Neither the period nor totalize operating modes are particularly important in communications troubleshooting.

All electronic counters have some form of *main gate* that controls the count start and stop with respect to time. Usually, the main gate is a form of AND gate. All electronic counters have some form of *time base* that supplies the precise increment of time to control the gate for a frequency or pulse-train measurement. Usually, the time base is a crystal-controlled oscillator. The accuracy of the counter depends on the accuracy of the time base, plus or minus

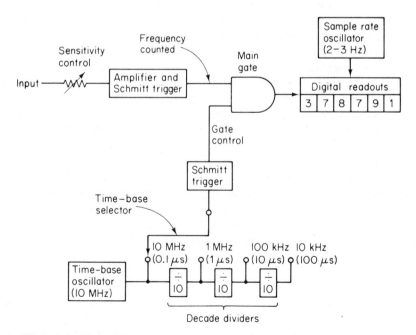

Figure 2-9 Basic digital counter circuit for frequency measurement operation.

one count. For example, if the time-base accuracy is 0.005%, the overall accuracy of the counter is 0.005%, plus or minus one count. The one-count error arises because the count may start and stop in the middle of an input pulse, thus omitting the pulse from the total count. Or part of the pulse may pass through the gate before the gate closes, thus adding a pulse to the count.

Most electronic counters have *dividers* that permit variation of gate time. These dividers convert the fixed-frequency time base to several other frequencies. In addition to the four basic sections, most electronic counters have attenuator networks, amplifier and trigger circuits to shape a variety of input signals to a common form, and logic circuits to control operation of the instrument.

Electronic counters have some form of *counter and readout.* Early instruments used binary counters and readout tubes that converted the binary count to a decade readout. Such instruments have generally been replaced by decade counters that convert the count to binary-coded decimal (BCD) form, decoders that convert the BCD data to decade form (generally BCD-to-seven segment decoders), and readouts that display the decade information (generally seven-segment LEDs, LCDs, etc.). One readout or display is provided for each digit. For example, eight readouts provide for a count up to 99,999,999.

Frequency measurement operation.   For frequency measurements, the digital counter circuits are arranged as shown in Fig. 2–9. The input signal (say from the transmitter output) is first converted to uniform pulses by the Schmitt trigger. These pulses are then routed through the main gate and into the counter/readout circuits, where the pulses are totalized. The number of pulses totalized during the "gate-open" interval is a measure of the average input frequency for that interval. For example, assume that the gate is held open for 1 and the count is 333. This indicates a frequency of 333 Hz. The count obtained is then displayed (with the correct decimal point) and retained until a new sample is ready to be shown. The sample rate oscillator determines the time between samples (not the interval of gate opening and closing), resets the counter, and thus initiates the next measurement cycle.

The time-base selector switch selects the gating interval, thus positioning the decimal point and selecting the appropriate measurement units. The time-base selector selects one of the frequencies from the time-base oscillator. If the 10-MHz signal (directly from the time base shown in Fig. 2–9) is selected, the time interval (gate-open to gate-close) is $1/10$ μs. If the 1-MHz signal (from the first decade divider in Fig. 2–9) is chosen, the measurement time interval is 1 μs.

Counter accuracy.   As discussed, the accuracy of a frequency counter is set by the stability of the time base rather than the readout. The readout is typically accurate to within ± 1 count. The time base of the Fig. 2–9 counter is 10 MHz and is stable to within ± 10 ppm (parts per million), or 100 Hz. The time base of a precision laboratory counter could be on the order of 4 MHz and is stable to within ± 1 ppm, or 4 Hz.

Counter resolution.   The resolution of an electronic counter is set by the number of digits in the readout. For example, assume that you must use a five-digit counter to troubleshoot the circuits of a CB set. The CB operating frequencies or channels are in the 27-MHz range. Now assume that you measure a 27-MHz signal with the five-digit counter. The count could be 26.999 or 27.001, or within 1000 Hz of 27 MHz. Since the FCC requires that the operating frequency of a CB set be held within 0.005% (or about 1350 Hz in the case of a 27-MHz signal), a digital counter for CB troubleshooting must have a minimum of five digits in the readout.

Combining accuracy and resolution.   To find out if a counter is adequate for a particular communications test, add the time-base stability (in terms of frequency) to the resolution at the operating frequency. Again using the CB set example, if the accuracy is 100 Hz, and the count can be resolved to 1000 Hz (at the measurement frequency), the maximum possible inaccuracy is 1000 + 100 Hz, or 1100 Hz. This is within the approximate 1350 Hz (0.005% of 27 MHz) required.

2-5.3 Calibration Check of Frequency Meters and Counters

The accuracy of frequency-measuring devices (both meters and counters) used for communications troubleshooting should be checked periodically, at least every 6 months. Always follow the procedures recommended in the frequency meter or counter service instructions. Generally, you can send the instrument to a calibration lab, or to the factory, or you can maintain your own frequency standard. (This latter is generally not practical for most communications service shops.)

No matter what standard is used, keep in mind that the standard must be more accurate, and have better resolution, than the frequency-measuring device, just as the meter or counter must be more accurate than the communications equipment.

2-5.4 Using WWV Signals for Frequency Calibration

In the absence of a frequency standard, or factory calibration, you can use the frequency information broadcast by U.S. government radio station WWV. These WWV signals are broadcast on 2.5, 5, 10, 15, 20, and 25 MHz continuously night and day, except for silent periods of approximately 4 min beginning 45 min after each hour. Broadcast frequencies are held accurate to within 5 parts in 10^{11}. This is far more accurate than that required for most communications equipment tests.

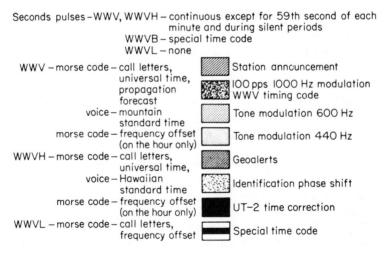

Seconds pulses – WWV, WWVH – continuous except for 59th second of each
                                    minute and during silent periods
                        WWVB – special time code
                        WWVL – none

Figure 2-10   Hourly broadcast schedules of WWV.

The hourly broadcast schedules of WWV are shown in Fig. 2–10. However, these schedules are subject to change. For full data on WWV broadcasts, refer to NBS (National Bureau of Standards) Standard Frequency and Time Services (Miscellaneous Publication 236), available from the Superintendent of Documents, U.S. Government Printing Office, Washington, D.C. 20402.

It is the continuous-wave (CW) signals broadcast by WWV that provide the most accurate means of calibrating (or checking) frequency meters and counters. It is not practical to use the signal directly, except on some special frequency meters, but the test connections for check are not complex.

Figure 2–11 shows the basic test connections for checking the accuracy of a frequency counter using WWV. Note that a receiver and signal generator are required. The accuracy of the signal generator and receiver are not critical, but both instruments must be capable of covering the desired frequency range. The procedure is as follows:

1. Allow the signal generator, receiver, and counter being tested to warm up for at least 15 min.

2. Reduce the signal generator output amplitude to zero. Turn off the signal generator output if this is possible without turning off the entire signal generator.

3. Tune the receiver to the desired WWV frequency. It is generally best to use a WWV frequency that is near the operating frequency of the communications equipment. For example, if a 27-MHz CB set is being tested, use the 25-MHz WWV signal.

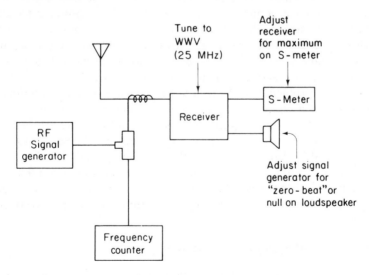

Figure 2–11   Basic test connections for checking the accuracy of a frequency counter using WWV.

4. Operate the receiver controls until you can hear the WWV signal in the receiver loudspeaker.

5. If the receiver is of the communications type, it will have a beat-frequency oscillator (BFO) and an output signal strength or S-meter. Turn on the BFO, if necessary, to locate and identify the WWV signal. Then tune the receiver for maximum signal strength on the S-meter. The receiver is now exactly on 25 MHz, or whatever WWV frequency is selected.

6. Turn on the signal generator, and tune the generator until it is at "zero beat" against the WWV signal. As the signal generator is adjusted so that its frequency is close to that of the WWV signal (so that the difference in frequency is within the audio range), a tone, whistle, or "beat note" is heard on the receiver. When the signal generator is adjusted to exactly the WWV frequency, there is no "difference signal," and the tone can no longer be heard. In effect, the tone drops to zero, and the two signals (generator and WWV) are at "zero beat."

7. Read the counter. The readout should be equal to the WWV frequency. For example, with a five-digit counter at 25 MHz, the reading should be 24.999 to 25.001.

8. Repeat the procedure at other WWV broadcast frequencies.

2-6. DUMMY LOAD

Never adjust a radio transmitter without an antenna or load connected to the output. This will almost certainly cause damage to the transmitter circuits. When a transmitter is connected to an antenna or load, power is transferred from the final RF stage to the antenna or load. Without an antenna or load, the final RF stage must dissipate the full power and will probably be damaged. Equally important, you should not take any major adjustments to a transmitter that is connected to a radiating antenna. You will probably cause interference.

These two problems can be overcome by means of a nonradiating load, commonly called a *dummy load*. There are a number of commercial dummy loads for communications equipment troubleshooting. The RF wattmeters described in Sec. 2-7 and the special test sets covered in Sec. 2-13 contain dummy loads. It is also possible to make up dummy loads suitable for most communications troubleshooting. There are two generally accepted dummy loads: the fixed resistance and the lamp. Keep in mind that these loads are for routine troubleshooting; they are not a substitute for an RF wattmeter or special test set.

2-6.1 Fixed-Resistor Dummy Load

The simplest dummy load is a fixed resistor capable of dissipating the full power output of the transmitter. The resistor can be connected to the transmitter antenna connector by means of a plug, as shown in Fig. 2-12.

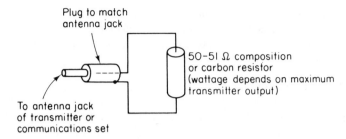

Plug to match
antenna jack

50-51 Ω composition
or carbon resistor
(wattage depends on maximum
transmitter output)

To antenna jack
of transmitter or
communications set

Figure 2-12   Fixed-resistor dummy load.

Most communications transmitters operate with a 50-Ω antenna and lead-in, and thus require a 50-Ω resistor. The nearest standard resistor is 51 Ω. This 1-Ω difference is not critical. However, it is essential that the resistor be noninductive (composition or carbon), never wire-wound. Wire-wound resistors have some inductance, which changes with frequency. Thus, the load (impedance) presented by the wire-wound resistor changes with frequency.

Always use a resistor with a power rating greater than the anticipated maximum output power of the transmitter. For example, an AM CB transmitter can (legally) have a 5-W input, which results in an output of about 4 W. A 7- to 10-W resistor should be used for the dummy load.

RF power output measurement with a dummy-load resistor.   It is possible to get an approximate measurement of RF power output from a radio transmitter with a resistor dummy load and a suitable meter. Again, these procedures are not to be considered a substitute for power measurement with an accurate RF wattmeter.

The procedure is simple. Measure the voltage across the 50-Ω dummy-load resistor and find the power with the equation

$$\text{power} = \frac{(\text{voltage})^2}{50}$$

For example, if the voltage measured is 14 V, the power output is

$$\frac{(14)^2}{50} = 3.92 \text{ W}$$

which is typical for an AM CB transmitter.

Certain precautions must be observed. First, the meter must be capable of producing accurate voltage indications at the transmitter operating frequency. This usually requires a meter with an RF probe (preferably a probe calibrated with the meter). An AM or FM transmitter should be checked with an RMS voltmeter and with no modulation applied. An SSB transmitter must be checked with a peak-reading voltmeter, and with modulation applied (since SSB produces no output without modulation). This usually involves connecting an audio generator to the microphone input of the SSB transmitter circuits. Always follow the service literature recommendations for all RF power output measurements (frequency, channels, operating voltages, modulation, etc.).

2-6.2 Lamp Dummy Load

Lamps have been the traditional dummy loads for communications equipment troubleshooting. For example, the No. 47 lamp (often found as a pilot lamp in many electronic instruments) provides the approximate impedance and power dissipation required as a dummy load for CB equipment. The connections are shown in Fig. 2–13.

You cannot get an accurate measurement of RF power output when a lamp is used as the dummy load. However, the lamp provides an indication of the relative power and shows the presence of modulation. The intensity of the light produced by the lamp varies with modulation (more modulation, brighter glow). Thus, you can tell at a glance if the transmitter is producing an RF carrier (steady glow), and if modulation is present (varying glow).

2-7. RF WATTMETER

A number of commercial RF wattmeters are available for communications troubleshooting. Also, the special test sets described in Sec. 2–14 usually include an RF wattmeter. The basic RF wattmeter consists of a dummy load (fixed resistor) and a meter that measures voltage across the load, but reads out in watts (rather than in volts), as shown in Fig. 2–14. You simply connect the RF wattmeter to the antenna connector of the set (transmitter output), key the transmitter, and read the power output on the wattmeter scale.

Although operation is simple, you must remember that SSB transmitters require a peak-reading wattmeter to indicate PEP (peak envelope power), whereas an AM or FM set uses an RMS-reading wattmeter. Most commercial RF wattmeters are RMS-reading, unless specifically designed for SSB.

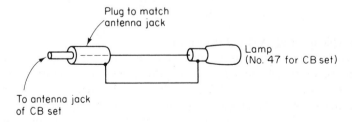

Figure 2–13   Lamp dummy load.

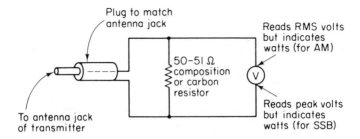

Figure 2-14   Basic RF wattmeter circuit.

2-8. FIELD STRENGTH METER

There are two basic types of field strength meters: the simple relative field strength (RFS) meter and the precision laboratory or broadcast-type instrument. Most communications equipment troubleshooting can be carried out with simple RFS instruments. An exception is where you must make precision measurements of broadcast antenna radiation patterns.

The purpose of a field strength meter is to measure the strength of signals radiated by an antenna. This simultaneously tests the transmitter output, the antenna, and the lead-in. In the simplest form, a field strength meter consists of an antenna (a short piece of wire or rod), a potentiometer, diodes, and a microammeter, as shown in Fig. 2-15. More elaborate RFS meters include a tuned circuit and possibly a transistor amplifier. In use, the meter is placed near the antenna at some location accessible to the transmitter or set (where you can see the meter), the transmitter is keyed, and the relative field strength is indicated on the meter. Some of the special test sets described in Sec. 2-13 include an RFS test function.

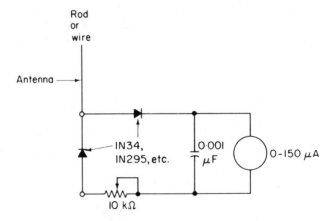

Figure 2-15   Basic relative-field-strength (RFS) meter circuit.

2-9. STANDING-WAVE-RATIO MEASUREMENT

The standing-wave ratio of an antenna is actually a measure of match or mismatch for the antenna, transmission line (lead-in), and the communications set. When the impedances of the antenna, line, and set are perfectly matched, all the energy or signal is transferred to or from the antenna, and there is no loss. If there is a mismatch (as is the case in any practical application), some of the energy or signal is reflected back into the line. This energy cancels part of the desired signal.

If the voltage (or current) is measured along the line, there are voltage or current maximums (where the reflected signals are in phase with the outgoing signals), and voltage or current minimums (where the reflected signal is out of phase, partially canceling the outgoing signal). The maximums and minimums are called *standing waves*. The ratio of the maximum to the minimum is the standing-wave ratio (SWR). The ratio may be related to either voltage or current. Since voltage is easier to measure, it is usually used, resulting in the common term *voltage standing-wave ratio* (VSWR). The theoretical calculations for VSWR are shown in Fig. 2–16.

An SWR of 1-to-1, expressed as 1:1, means that there are no maximums or minimums (the voltage is constant at any point along the line) and that there is a perfect match for set, line, and antenna. As a practical matter, if this 1:1 ratio should occur on one frequency, it will not occur at any other frequency, since impedance changes with frequency. It is not likely that all three elements (set, antenna, line) will change impedance by exactly the same amount on all frequencies. Therefore, when checking SWR, always check on all frequencies or channels, where practical. As an alternative, check SWR at the high, low, and middle channels or frequencies.

In the case of microwave signals being measured in the laboratory, a meter is physically moved along the line to measure maximum and minimum voltages. This is not practical at most communications equipment frequencies, due to the physical length of the waves. In communications equipment, it is far

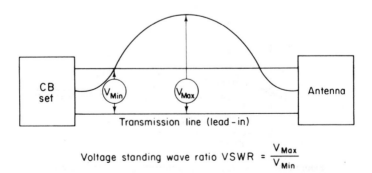

$$\text{Voltage standing wave ratio VSWR} = \frac{V_{Max}}{V_{Min}}$$

Figure 2–16   Calculations for voltage standing-wave ratio (VSWR).

more practical to measure forward and outgoing voltage and reflected voltage, and then calculate the *reflection coefficient* (reflected voltage/outgoing voltage). The relationship of reflection coefficient to SWR is as follows:

$$\text{reflection coefficient} = \frac{\text{reflected voltage}}{\text{forward voltage}}$$

For example, using a 10-V forward and a 2-V reflected voltage, the reflection coefficient is 0.2.

Reflection coefficient is converted to SWR by dividing (1 + reflection coefficient) by (1 − reflection coefficient). For example, using the 0.2 reflection coefficient, the SWR is

$$\frac{1 + 0.2}{1 - 0.2} = \frac{1.2}{0.8} = 1.5 \text{ SWR}$$

This may be expressed as 1 to 1.5, 1.5 to 1, 1.5:1, 1:1.5, or simply as 1.5, depending on the meter scale. In practical terms, an SWR of 1.5 is poor, since it means that at least 20% of the power is being reflected.

SWR can be converted to reflection coefficient by dividing (SWR − 1) by (SWR + 1). For example, using 1.5 SWR, the reflection coefficient is

$$\frac{1.5 - 1}{1.5 + 1} = \frac{0.5}{2.5} = 0.2 \text{ reflection coefficient}$$

In the commercial SWR meters used for communications work, it is not necessary to calculate either reflection coefficient or SWR. This is done automatically by the SWR meter. (The meter is actually reading the reflection coefficient, but the scale reads in SWR. If you have a reflection coefficient of 0.2, the SWR reading is 1.5.)

There are a number of SWR meters used in communications work. Some communications sets, such as most CB sets, have built-in SWR meters and circuits. The SWR function is often combined with other measurement functions (field strength, power output, etc.). Practically all of the special test sets described in Sec. 2-13 include an SWR measurement feature, since it is so important to proper operating of communication sets.

Basic SWR meter circuits are quite simple, and it is possible to build them in the shop. However, it is not practical in most cases to do so. The basic circuit requires that a *directional coupler* be inserted in the transmission line. Even under good conditions, a mismatch and some power loss may result. A poorly designed pickup may result in considerable power loss, as well as inaccurate readings. Thus, it is more practical to use commercial SWR meters.

The basic SWR meter circuit is shown in Fig. 2-17. Operation of the circuit is as follows. As shown, there are two pickup wires, both parallel to the center conductor of the transmission line. Any RF voltage on either of the parallel pickups is rectified and applied to the meter through switch S1. Each

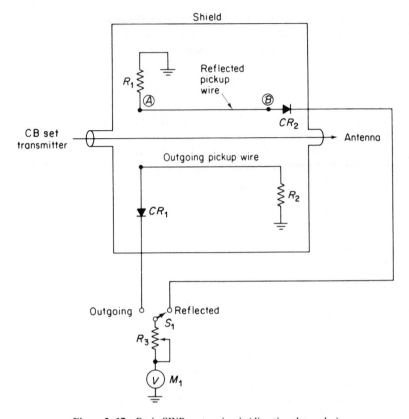

Figure 2-17 Basic SWR meter circuit (directional coupler).

pickup wire is terminated in the impedance of the transmission line by corresponding resistors R1 and R2 (typically 50 to 52 Ω).

The outgoing signal (transmitter to antenna) is absorbed by R1. Thus, there is no outgoing voltage on the reflected pickup wire beyond point A. However, the outgoing voltage remains on the transmission line at the outgoing pickup wire. This signal is rectified by CR1, and appears as a reading on the meter, when S1 is in the outgoing voltage position.

The opposite condition occurs for the reflected voltage (antenna to transmitter). There is no reflected voltage on the outgoing pickup wire beyond point B, because the reflected voltage is absorbed by R2. The reflected voltage does appear on the reflected pickup wire beyond this point and is rectified by CR2. The reflected voltage appears on meter M1 when S1 is in the reflected voltage position.

In use, switch S1 is set to read outgoing voltage, and resistor R3 is adjusted until the meter needle is aligned with some "set" or "calibrate" line (near the right-hand end of the meter scale). Switch S1 is then set to read reflected voltage, and the meter needle moves to the SWR position.

As a practical matter, SWR meters often do not read beyond 1:3. This is because an indication above 1:3 indicates a poor match. Make certain that you understand the scale used on the SWR meter. For example, a typical SWR meter is rated at 1:3, meaning that it reads from 1:1 (perfect) to 1:3 (poor). However, the scale indications are 1, 1.5, 2, and 3. These scale indications mean 1:1, 1:1.5, 1:2, and 1:3, respectively. The scale indications between 1 and 1.5 are the most useful, since a good antenna system shows a typical 1.1 or 1.2. Anything between 1.2 and 1.5 is on the borderline.

2-10. DIP METERS

The dip meter, or grid-dip adapter, has long been a tool in radio communications service work, particularly in amateur radio. The dip meter has many uses, but its most useful function in communications work is in presetting "cold" transmitter and receiver resonant circuits (no power applied to the set). This makes it possible to adjust the resonant circuits of a badly tuned set, or a set where new coils and transformers must be installed as a replacement.

As an example, it is possible that the replacement coil or transformer is tuned to an undesired frequency when shipped from the factory. Using a dip meter, it is possible to install the coil, tune it to the correct frequency, and then apply power to the set and adjust the circuit for "peak" as described in the service literature. (Most service literature assumes that the circuits are not badly tuned or only require "peaking.")

There are many types of dip meters and circuits. A typical dip meter is a hand-held, battery-operated device. The circuit is essentially an RF oscillator with external coil, a tuning dial, and a meter. When the coil is held near the circuit to be tested, and the oscillator is tuned to the resonant frequency of the test circuit, part of the RF energy is absorbed by the test circuit, and the meter indication "dips." The procedure can be reversed, where the dip meter is set to a desired frequency and the test circuit is tuned to produce a "dip" indication on the meter. We will not go into the many uses of the dip meter here. Instead, we shall describe how a dip adapter may be used to preset resonant circuits or to check the frequency of resonant circuits.

2-10.1 Basic Dip Adapter

A basic dip adapter circuit is shown in Fig. 2-18. Such a circuit may be fabricated in the shop with little difficulty. Resistor R1 should match the impedance of the signal generator (typically 50 Ω). Both diode CR1 and the microammeter should match the output of the signal generator. The pickup coil L1 consists of a few turns of insulated wire. The accuracy of the dip adapter circuit depends on the counter accuracy, or on the signal generator accuracy if the counter is omitted.

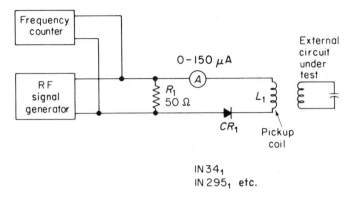

Figure 2-18   Basic dip-adapter circuit.

2-10.2 Setting Resonant Frequency with a Dip Adapter

The frequency of a resonant circuit may be set using a dip adapter. The following procedure is applicable to both series and parallel resonant circuits.

1. Couple the dip adapter to the resonant circuit using pickup coil L1 of Fig. 2-18. Usually, the best coupling has a few turns of L1 passed over the coil of the resonant circuit. Make certain that the communications set is off.

2. Set the signal generator to the desired resonant frequency, as indicated by the frequency counter. Adjust the signal generator output amplitude control for a convenient reading on the adapter meter.

3. Tune the resonant circuit for a maximum dip on the adapter meter. The resonant circuits of the set may be tuned by means of adjustable slugs in the coil and/or adjustable capacitors.

4. Most resonant circuits are designed so as not to tune across both the fundamental frequency and any harmonics. However, it is possible that the circuit will tune to a harmonic and produce a dip. To check this condition, tune the resonant circuit for maximum dip and set the signal generator to the first harmonic (twice the desired resonant frequency) and to the first subharmonic (one-half the resonant frequency). Note the amount of dip at both harmonics. The harmonics should produce substantially smaller dips than the fundamental resonant frequency.

5. For maximum accuracy, check the dip frequency from both high and low sides of the resonant circuit tuning. A significant difference in frequency readout from either side indicates overcoupling between the dip adapter circuit and the resonant circuit under test. Move the adapter coil L1 away from the test circuit until the dip indication is just visible. This amount of coupling should provide maximum accuracy. (If there is difficulty in finding a dip, overcouple the adapter until a dip is found, then loosen the coupling and make a final check of frequency. Generally, the dip is more pronounced when it is approached from the direction that causes the meter reading to rise.)

6. If there is doubt as to whether the adapter is measuring the resonant frequency of the desired circuit or some nearby circuit, ground the circuit under test. If there is no change in the adapter dip reading, the resonance of another circuit is being measured.

7. The area surrounding the circuit being measured should be free of wiring scraps, solder drippings, and so on, as the resonant circuit can be affected by them (especially at high frequencies), resulting in inaccurate frequency readings. Keep fingers and hands as far away as possible from the adapter coil (to avoid adding body capacitance to the circuit under test).

8. All other factors being equal, the nature of a dip indication provides an approximate indication of the test circuit's Q factor. Generally, a sharp dip indicates a high Q, whereas a broad dip shows a low Q.

9. The dip adapter may also be used to measure the frequency to which a resonant circuit is tuned. The procedure is essentially the same as that for presetting the resonant frequency (steps 1 through 8), except that the signal generator is tuned for a maximum dip (communications set still cold and the test circuit untouched). The resonant frequency to which the test circuit is tuned is then read from the counter. When making this test, watch for harmonics, which also produce dip indications.

2-11. SPECTRUM ANALYZER AND FM DEVIATION METER

Communications waveforms (particularly those resulting from frequency modulation, or FM) are sometimes measured by means of a spectrum analyzer, especially in the lab. The FM deviation meter is generally more practical for everyday communications troubleshooting in the shop. Needless to say, an FM deviation meter is far less expensive than a spectrum analyzer. (FM deviation meters are readily available in kit form, such as the Heathkit IM-4180.) However, a discussion of spectrum analyzers will help you to understand the nature of communications waveforms, particularly FM waveforms.

2-11.1 Basic Spectrum Analyzer

The basic circuit of a spectrum analyzer is shown in Fig. 2-19. The spectrum analyzer is essentially a narrowband receiver, electrically tuned over a given frequency range, combined with an oscilloscope. As shown, the local oscillator is swept over a range of frequencies by a sweep generator circuit. Since the IF amplifier passband remains fixed, the input circuits and mixer are swept over a corresponding range of frequencies. For example, if the intermediate frequency is 10 kHz and the local oscillator sweeps a band of frequencies from 100 to 200 kHz, the input is capable of receiving signals in the

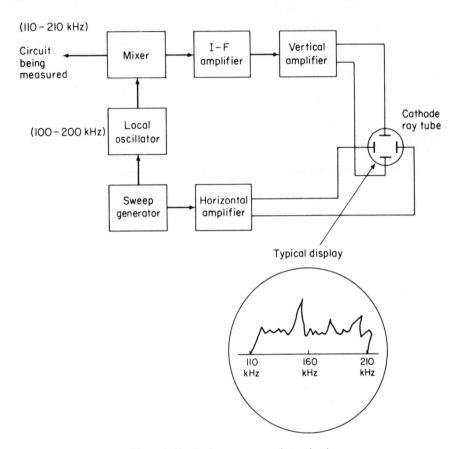

Figure 2-19   Basic spectrum analyzer circuit.

range from 110 to 210 kHz. The output of the IF amplifier is further amplified and supplied to the vertical deflection plates of a cathode-ray tube.

The cathode-ray-tube horizontal plates receive their signal from the same sweep generator used to tune the local oscillator. Thus, the length of the horizontal sweep represents the total sweep spectrum. For example, if the sweep is from 110 to 210 kHz, the left-hand end of the horizontal trace represents 110 kHz and the right-hand end represents 210 kHz, with the midpoint of the trace representing 160 kHz, as shown in Fig. 2-19.

Spectrum analyzers are often used in conjunction with Fourier analysis and transform analysis. Both of the techniques are quite complex and beyond the scope of this book. Instead, we concentrate on the practical aspects of spectrum analysis during communications troubleshooting. That is, we discuss what display results from a given input signal, and how the display can be interpreted.

2–11.2 Unmodulated Signal Displays

If the spectrum analyzer's local oscillator sweeps slowly through an un-modulated or continuous-wave (CW) signal, the resulting response on the analyzer screen is simply a plot of the analyzer's IF amplifier passband. A pure CW signal has, by definition, energy at only one frequency, and should therefore appear as a single spike on the analyzer screen. This occurs provided that the total sweep width or so-called "spectrum width" is wide enough com-pared to the IF bandwidth in the analyzer. As spectrum width is reduced, the spike response begins to spread out until the IF bandpass characteristic begins to appear.

2–11.3 Amplitude-Modulated Signal Displays

A pure sine wave represents a signal frequency. The spectrum of a pure sine wave is shown in Fig. 2–20. Note that this is the same as an unmodulated signal display. The height of line F_0 represents the power contained in the single

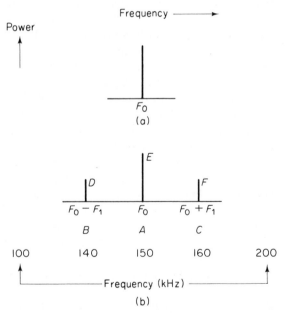

Position of A = carrier frequency (150 kHz)
Distance between B and A, or A and C = frequency modulation frequency (10 kHz)
Ratio of D to E, or F to E = one–half percentage of modulation

Figure 2-20 Frequency spectrum for single-tone amplitude-modulated carrier, and rules for interpretation: (a) unmodulated; (b) amplitude modulated with single fre-quency.

frequency. Figure 2–20(b) shows the spectrum for a single sine-wave frequency F_0, amplitude-modulated by a second sine wave F_1. In this case, two sidebands are formed, one higher than and one lower than the frequency F_0. These sidebands correspond to the sum and difference frequencies, as shown. If more than one modulating frequency is used (as is the case with most practical AM signals), two sidebands are added for each frequency.

Note that if the frequency, spectrum width, and vertical response of the analyzer are calibrated (as they are with any modern laboratory instrument) it is possible to find (1) the carrier frequency, (2) the modulation frequency, (3) the modulation percentage, and (4) the nonlinear modulation (if any) and incidental FM (if any).

An amplitude-modulated spectrum display can be interpreted as follows:

The *carrier frequency* is determined by the position of the center vertical line F_0 on the X axis. For example, if the total spectrum is from 100 to 200 kHz and F_0 is in the center as shown in Fig. 2–20(b), the carrier frequency is 150 kHz.

The *modulation frequency* is determined by the position of the sideband lines $F_0 - F_1$ or $F_0 + F_1$ on the X axis. For example, if sideband $F_0 - F_1$ is at 140 kHz, and F_0 is at 150 kHz as shown, the modulating frequency is 10 kHz. Under these conditions, the upper sideband $F_0 + F_1$ should be 160 kHz. The distance between the carrier line F_0 and either sideband is sometimes known as the *frequency dispersion* and is equal to the modulation frequency.

The *modulation percentage* is determined by the ratio of the sideband amplitude to the carrier amplitude. The amplitude of either sideband with respect to the carrier voltage is one-half of the percentage of modulation. For example, if the carrier amplitude is 100 mV and either sideband is 50 mV, this indicates 100% modulation. If the carrier amplitude is 100 mV and either sideband is 33 mV, this indicates 66% modulation.

Nonlinear modulation is indicated when the sidebands are of unequal amplitude or are not equally spaced on both sides of the carrier frequency. Unequal amplitude indicates nonlinear modulation that results from a form of undesired frequency modulation combined with the amplitude modulation.

Incidental FM is indicated by a shift in the vertical signals along the X axis. For example, any horizontal "jitter" of the signals indicates rapid frequency modulation of the carrier.

The rules for interpreting AM spectrum analyzer displays are summarized in Fig. 2–20.

In practical tests, carrier signals are often amplitude-modulated at many frequencies simultaneously. This results in many sidebands (two for each modulating frequency) on the display. To resolve this complex spectrum, the operator should make sure that the analyzer bandwidth is less than the lowest modulating frequency or less than the difference between any two modulating frequencies, whichever is smaller.

Overmodulation also produces extra sideband frequencies. The spectrum for overmodulation is very similar to multifrequency modulation. However, overmodulation is usually distinguished from multifrequency modulation by the facts that (1) the spacing between overmodulated sidebands is equal, while multifrequency sidebands may be arbitrarily spaced (unless the modulating frequencies are harmonically related); and (2) the amplitude of the overmodulated sidebands decreases progressively out from the carrier, but the amplitude of multifrequency-modulated signals is determined by the modulation percentage of each frequency and can be arbitrary.

2-11.4 Frequency-Modulated Signal Displays

The mathematical expression for a frequency-modulated waveform is long and complex, involving a special mathematical operator known as a *Bessel function*. However, the spectrum representation of the FM waveform is straightforward. Such a representation is shown in Fig. 2-21, which illustrates

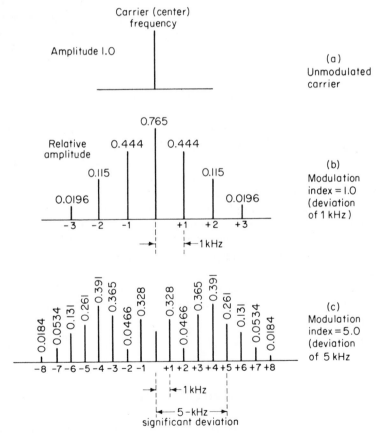

Figure 2-21   Frequency spectrum for single-tone (1-kHz) frequency-modulated carrier.

the frequency spectrum of a carrier that has been frequency-modulated by a single (1-kHz) sine wave. Figure 2–21 (a) shows the unmodulated carrier spectrum waveform. Figure 2–21(b) shows the relative amplitudes of the waveform when the carrier is frequency-modulated with a deviation of 1 kHz (modulation index of 1.0). Figure 2–21(c) shows the relative amplitudes of the waveform when the carrier is frequency-modulated with a deviation of 5 kHz (modulation index of 5.0). Note that the modulation index is given by:

$$\text{modulation index} = \frac{\text{maximum frequency deviation}}{\text{modulating frequency}}$$

The term *maximum frequency deviation* is theoretical. If a CW signal F_C is frequency-modulated at a rate F_R, an infinite number of sidebands result. These sidebands are located at intervals of $F_C \pm N_F$, where N = 1, 2, 3, and so on.

However, as a practical matter, only the sidebands containing significant power are usually considered. For a quick approximation of the bandwidth occupied by the significant sidebands, multiply the sum of the carrier deviation and the modulating frequency by 2:

bandwidth = 2 (carrier deviation + modulating frequency)

As a guideline, when using a spectrum analyzer to find the maximum deviation of an FM signal, locate the sideband where the amplitude begins to drop and continues to drop as the frequency moves from the center. For example, in Fig. 2–21(c), sidebands 1, 2, 3, and 4 rise and fall, but sideband 5 falls, and all sidebands after 5 continue to fall. Since each sideband is 1 kHz from the center, this indicates a practical or significant deviation of 5 kHz. (It also indicates a modulation index of 5.0.)

As in the case of amplitude modulation, the center frequency and modulating frequency can be determined by the spectrum analyzer display.

The *carrier frequency* is determined by the position of the center vertical line on the X axis. (The centerline is not always the highest amplitude, as shown in Fig. 2–21.)

The *modulating frequency* is determined by the position of the sidebands in relation to the centerline or the distance between sidebands (frequency dispersion).

2–11.5 FM Deviation Meter

The operating controls and procedures for an FM deviation meter are far less complex than those of a spectrum analyzer. Typically, FM deviation meter controls include a meter scale marked off in terms of frequency, a tuning control, and a zero control. In use, the FM deviation meter is connected to monitor the output of the communications transmitter. The transmitter is first keyed without modulation so that the meter can be tuned to the exact carrier frequency. Then the transmitter is frequency-modulated with a tone (typically in the range 1 to 5 kHz, and the exact amount of frequency modulation is indicated on the FM deviation meter.

2-12. MISCELLANEOUS COMMUNICATIONS TROUBLESHOOTING EQUIPMENT

There are many items of equipment that will make life easier for the communications service technician, but they are not absolutely essential for communications troubleshooting. The following are some examples.

2-12.1 Base Station Set and Antenna

The uses of a known good communications set and base station (or shop) antenna are obvious. You can check the operation of a suspected set on the good shop antenna. If the set performs properly with the shop antenna, but not with its own antenna, the problem is localized. You can reverse the procedure and test the suspected antenna with a known good set. Also, you can communicate between the shop and another station (either mobile or base station) to check operation, before and after service.

Walkie-talkie CB.   A walkie-talkie CB may also be used for communications from the shop to remove locations for field-strength and other tests. Keep in mind that the walkie-talkie must be licensed under Part 95 of the FCC regulations if you communicate with other CB stations (unless you communicate only between unlicensed walkie-talkies). A walkie-talkie may also be used to track down electrical interference.

2-12.2 Power Supply and Isolation Transformer

A well-equipped communications service shop should have at least two power supplies: one a-c supply, variable from about 100 to 130 V ac; and one d-c supply, variable from about 10 to 15 V dc. There are a number of commercial power supplies that meet these requirements, so we shall not discuss the circuits.

Most commercial power supplies include a transformer. Often, this is a variable auto-transformer (or variac). The use of a transformer in the power supply eliminates the need for an *isolation transformer* as discussed in Sec. 1-7.

2-12.3 Distortion Meters

Some service shops include distortion meters or distortion analyzers. There are two basic types: the *intermodulation distortion analyzer* and the *harmonic distortion analyzer*. Before you rush to buy either of these instruments, consider the following. The basic purpose of any distortion meter or analyzer is to measure the amount of distortion (usually as a percentage), not to locate the cause of distortion. The circuits in a communications set where distortion meters may be used effectively are the audio/modulation circuits. These circuits are designed for operation at frequencies below about 3 kHz (typically). Distortion meters are usually used in audio service work, where frequencies are in the range dc to 20 kHz. Generally, if there is sufficient distortion in the audio/modulation circuits to be a problem, you will hear it in both transmission and reception.

2-12.4 Communications Receivers, Spectrum Analyzers, and TV Sets

In addition to checking that a communications set produces the correct signals on all channels, it is helpful to know that the set is not producing any other signals! For example, the final RF amplifier in the transmitter may break into oscillation (if not properly neutralized) and produce signals at undetermined frequencies. These may not show up on the channel being used, or on any other channel. The problem is most common in vacuum-tube sets. More likely, the transmitter may produce harmonics that interfere with television and with other radio communications services. (Television interference is a very common problem in CB communications.) All of these extra signals (generally referred to as *spurious signals* in FCC regulations and service literature) are illegal, and certainly undesirable.

The ideal instrument for detecting undesired signals from a communications set is the spectrum analyzer described in Sec. 2-11. Although the spectrum analyzer is ideal, it is also very expensive, and is therefore generally restricted to laboratory use or commercial broadcast work. You can do essentially the same job with a good communications-type receiver. The communications receiver should have a BFO (beat-frequency oscillator) as well as an S-meter (signal strength meter). In addition to using the communications receiver for signal checks, you can monitor transmissions of sets being serviced, and the receiver may be used for WWV checks as described in Sec. 2-5.4.

One of the most frequent types of interference caused by communications sets is on television channels (especially CB communications). A television set in the shop will quickly indicate whether a communications set being serviced is causing any interference. The shop TV set may help you settle some disputes concerning TV interference problems. However, before you become overconfident, keep the following in mind. Most interference enters the TV set through the IF amplifiers, and the IF amplifiers of all TV sets do not operate on the same frequency. Some TV sets use the range 22 to 28 MHz, whereas other sets use the range 41 to 47 MHz (and some very old TV sets use other IF ranges). So it is possible for a set to produce interference on one TV and not on another, with both TV sets located in the same room and tuned to the same channel!

2–13. SPECIAL TEST SETS FOR COMMUNICATIONS SERVICE

There are a number of test sets designed specifically for communications service. Some of the sets are for field use, whereas others are for the bench or shop. Still other sets may be used in either the shop or field. The following paragraphs describe one such special test set. Keep in mind that this is not the only test set available, now and in the future, but represents a typical test set that incorporates the most important required functions of a communications test set.

Figure 2–22 shows a multipurpose tester (the Pace P-5425) suitable for checking communications transceivers in the range 25 to 50 MHz. The instrument is particularly suited for testing mobile CB units. The tester will measure power outputs up to 25 W (or 250 W when an external dummy load is used), SWR up to 1:3, percent of modulation up to 100%, relative field strength, and crystal activity (on a good-bad basis). The tester also provides a built-in 25-W dummy load, a crystal-controlled RF oscillator at 27 MHz, and an audio oscillator at 1000 Hz (which can also be used to modulate the 27-MHz RF oscillator).

The same meter is used for all functions. Operation of the meter is controlled by the selector switches as follows:

In PWR, the meter is connected to read the forward voltage applied to the dummy load (25 W) or to an external load (up to 250 W).

In SWR, the meter is connected to read both forward and reflected voltages in the directional coupler, depending on the position of the SET–CHECK switch.

In MOD, the meter is connected to read both forward and reflected voltages in the directional coupler, depending on the position of the SET–CHECK switch. In the CHECK position, the meter is connected through the 2SB56 transistor circuit to read only the audio or modulation voltage of the RF carrier, as a percentage of modulation.

In RFS, the meter is connected to read the rectified signal present on a telescopic antenna connected to the RFS ANT jack. The rectified or detected signals are also available at the SCOPE terminal, permitting the audio or modulation signals to be displayed on an external oscilloscope.

In XTAL/RF OSC, the meter is connected to read a portion of the 9 V (from the tester's internal battery) applied to the RF oscillator. A crystal to be tested is inserted into the RF oscillator XTAL socket and the meter is adjusted to read full scale by potentiometer VR1. The crystal is then removed, and the meter needle or pointer drops back to some point less than full scale. If the

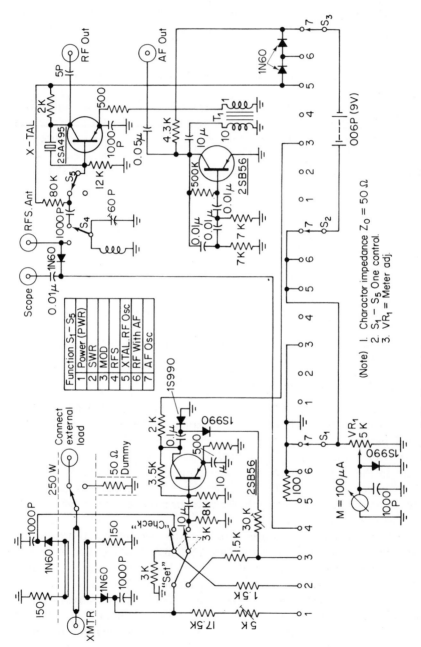

Figure 2-22 Pace P-5425 two-way test meter schematic.

94

pointer stops within the GOOD zone of the XTAL scale on the meter, the crystal under test is satisfactory for use in a communications set (but not necessarily on-frequency). If a defective crystal is tested, the RF oscillator does not oscillate, and the meter pointer remains in the BAD zone after the crystal is removed.

With a good crystal in the XTAL socket and the selector at XTAL/RF OSC, an unmodulated RF output is available from the RF OUT jack. This signal may be used to test operation of a communications receiver, or as a frequency standard (depending upon the accuracy of the crystal).

In RF WITH AF, the meter is grounded and produces no indication. Power (9 V) is applied to both the RF and AF oscillators. With a good crystal in the XTAL socket, an RF output (modulated at about 1000 Hz) is available from the RF OUT jack. A signal is also available from the AF OUT jack.

If the crystal used in the RF oscillator is at a frequency corresponding to a communications channel, the RF OUT signal may be used for signal injection to check operation of the receiver from antenna to loudspeaker. (The 1000-Hz tone should be heard in the loudspeaker.)

In AF OSC, the meter is grounded and produces no indication. Power (9 V) is applied to the AF oscillator, and an audio voltage at about 1000 Hz is available from the AF OUT jack. This audio signal may be used for signal injection to check operation of the receiver audio circuit (typically from detector or volume control to loudspeaker).

3

Communications
Troubleshooting

It is assumed that you are familiar with the basics of radio communications circuits (AM, FM, SSB), at least at a level found in the author's best-selling *Handbook of Practical CB Service* (Englewood Cliffs, N.J.: Prentice-Hall, Inc., 1978). The techniques found in this chapter are based on the troubleshooting methods covered in that book and in Chapter 1 of this book. However, the techniques here are advanced shortcuts designed to pinpoint quickly trouble in radio communications equipment to specific circuits and parts.

This chapter is divided into two parts. The first part describes a series of tests to check all sections of a typical communications set. The second part of the chapter tells how to use the test results to troubleshoot the various circuits. It is recommended that you perform both the test procedures and the troubleshooting steps in the order described in the following paragraphs (where practical). The sequence of test and troubleshooting steps is arranged to get the best possible shortcut for a typical radio communications set.

3-1. TRANSMITTER RF POWER CHECK

This is normally the first transmitter check performed. The test measures transmitter RF power to determine if the power level is normal. The check can be used for both AM and FM transmitters operated in the unmodulated condition. A similar test for SSB transmitters is described in Sec. 3–13.

NOTICE

FCC regulations require that all checks, adjustments, and repairs that affect transmitter power and frequency be performed only by or under the immediate supervision of persons holding a valid First or Second Class Radiotelephone License.

1. Connect the equipment as shown in Fig. 3-1. Make certain that the RF wattmeter (Sec. 2-7) is capable of measuring the maximum output power of the transmitter. If an RF wattmeter is not available, use a dummy load and meter to measure power as described in Sec. 2-6. If the transmitter is capable of both AM and FM, check the AM output power first (without modulation).

2. Set the transmitter to the first channel (or lowest frequency) to be checked. Operate the transmitter controls to produce the maximum output power.

3. Key the transmitter with the microphone push-to-talk switch (or whatever operating control is used). If necessary, cover the microphone so that the transmitter is not modulated.

4. Read transmitter RF power on the RF wattmeter (Sec. 2-7) or on the dummy load and RMS-reading meter (Sec. 2-6). As an example, most CB transmitters operate at the maximum allowable 5 W of input power, which results in an RF output power of 2.5 to 3.5 W. Refer to the manufacturer's specifications for normal RF output power of other transmitters.

5. Repeat steps 3 and 4 for each transmitter channel (or across the entire operating frequency range). Typically, RF power should be the same on all channels (or at all operating frequencies).

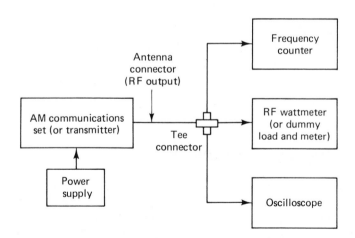

Figure 3-1   Test connections for transmitter RF power, frequency, and modulation checks.

6. Repeat steps 3 and 4 while applying some modulation (speak into the microphone). There should be substantially no change (theoretically) in RF power with modulation applied. As a practical matter, there may be some variation in RF output with modulation. (You will probably notice RF power output peaks when you speak loudly.) This condition is normal for most AM transmitters, but is highly undesirable for FM transmitters. As an example, FCC rules limit transmitter output power to 4 W under any condition of modulation when a CB transmitter is operated in the AM mode. Thus, a CB set that produces 3.5 W with no modulation can produce up to 4 W fully modulated, and still be considered as operating properly.

3-2. TRANSMITTER FREQUENCY CHECK

This check measures the accuracy of the transmitter operating frequency. The check should be performed simultaneously with the transmitter RF power check (Sec. 3-1). Immediately after reading the RF power from the wattmeter, read the transmitter frequency from the frequency counter. The check can be used for both AM and FM transmitters operated in the unmodulated condition. A similar test for SSB transmitters is described in Sec. 3-15.

1. Leave the equipment connected as shown in Fig. 3-1. Repeat steps 1 through 5 of the transmitter RF power checks (Sec. 3-1). Make certain that the transmitter is not modulated. If necessary, cover the microphone. If the microphone gain is adjustable, set for the lowest gain.

2. Read the transmitter frequency from the frequency counter (Sec. 2-5). Check all channels (or all operating frequencies). Note that if the transmitter shows no RF output, or the RF output is very low, you will probably not get an accurate frequency reading.

3. For an AM transmitter, repeat steps 1 and 2 while applying some modulation (speak into the microphone). In theory, there should be no change in frequency with modulation. Any substantial variation in frequency, with modulation, indicates problems, since the upper sidebands should cancel the lower sidebands.

3-3. TRANSMITTER MODULATION CHECKS

These checks show whether or not transmitter modulation is normal. The check for AM transmitters (Sec. 3-3.1) shows modulation by displaying the modulation envelope on an oscilloscope as described in Sec. 2-2. FM transmitters are checked by measuring output frequency deviation with fixed modulating input frequencies using an FM deviation meter (or possibly a spectrum analyzer). The checks should be performed after the transmitter power output and frequency have been checked. A similar test for SSB transmitters is described in Sec. 3-14.

3-3.1 AM Transmitter Modulation Check

1. Connect the equipment as described in Sec. 2-2. Use the modulation measurement method best suited to the available test equipment, following the notes and techniques discussed in Sec. 2-2.2 through 2-2.8.

2. There are two alternatives for connecting the audio modulation signal source (audio generator, Sec. 2-1.4) to the transmitter. You can connect the modulation source directly to the transmitter at the microphone input (or to a modulation input jack if the transmitter is so equipped). This provides the most stable and uniform modulation source. However, direct connection does not provide a check of the microphone. Also, on some transmitters, the set must be keyed by a push-to-talk switch on the microphone. The modulation audio generator can be connected to a loudspeaker. In turn, the microphone is placed directly over the loudspeaker. Use whichever method is most convenient.

3. If the transmitter is equipped with adjustable microphone gain, set the gain to midposition.

4. Secure the microphone over the speaker so that the speaker output drives the microphone with a constant tone. Adjust the modulation audio generator to a frequency of 1 kHz, unless otherwise specified by the transmitter test instructions.

5. Key the transmitter with the microphone push-to-talk switch. Adjust the oscilloscope for a stable display of the modulation envelope as described in Sec. 2-2.

6. Vary the gain control of the audio generator as necessary to obtain an oscilloscope display that shows modulation from 0 to 100%. If desired, the modulation nomogram of Fig. 2-6 and the techniques of Sec. 2-2.8 can be used to find the precise percentage of modulation.

7. Return the gain control of the audio generator to midposition. If the transmitter is equipped with adjustable microphone gain, vary the microphone gain from minimum to maximum setting. The observed modulation percentage should vary as the microphone gain control is adjusted.

8. Normally, the modulation check is necessary on only one channel. However, if a complete check is desired, leave the equipment set up as described in step 6 and adjust for 50% modulation. Select each channel in turn, and observe the oscilloscope display for any change. Unkey the transmitter while changing channels.

3-3.2 FM Transmitter Modulation Check

1. Connect the equipment as shown in Fig. 3-2. Make certain that the RF wattmeter or dummy load is capable of handling the maximum output power of the transmitter.

2. Set the transmitter to the first channel (or lowest frequency) to be

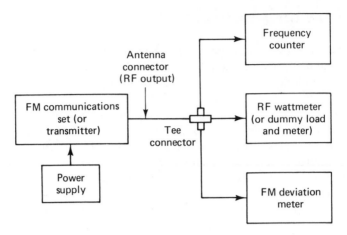

Figure 3-2   Test connections for FM transmitter RF power, frequency, and modulation checks.

checked. Operate the transmitter controls to produce the maximum output power (unmodulated).

3. Connect the audio modulation source to the transmitter using the most convenient alternative described in step 2 of Sec. 3-3.1.

4. Key the transmitter, without modulation, so that the FM deviation meter can be tuned to the exact carrier frequency.

5. If the transmitter is equipped with adjustable microphone gain, set the gain to midposition. Secure the microphone over the speaker so that the speaker output drives the microphone with a constant tone. Adjust the modulation audio generator to a frequency of 1 kHz, unless otherwise specified.

6. Key the transmitter and read the amount of deviation on the FM deviation meter. The indicated deviation should be equal to the frequency of the audio generator (1 kHz).

7. If desired, repeat the procedure using other audio generator (modulation) frequencies. A typical FM deviation meter will read up to about 7 or 8 kHz. Typical FM modulation test frequencies are 1 and 5 kHz.

8. If desired, repeat the tests on all channels.

3-4. RECEIVER AUDIO POWER CHECK

This is normally the first receiver check performed. The check quickly determines if the receiver is totally dead. If not, the check determines whether the audio section of the receiver can deliver adequate audio power. The check also accomplishes all the preliminary setup steps necessary for a receiver sensitivity check, which should be performed next if the results of this test are satisfactory. The test is designed primarily for AM receivers, but can also be used for FM communications (provided that an FM signal generator is available). The last part of the test also provides a quick check of the audio-frequency response by comparing the output power at 400 Hz and 2.5 kHz to the 1-kHz reference. This is a good overall check of the receiver's ability to pass all audio signals in the voice communications range. A similar test for SSB receivers is given in Sec. 3-16.

1. Connect the equipment as shown in Fig. 3-3. If the receiver loudspeaker must be disconnected, make certain to connect an audio load in place of the loudspeaker. Operation of the receiver without a loudspeaker or audio load will probably result in damage to the final audio transistors and circuits. The audio load should be a noninductive (not wire-wound) composition or carbon resistor of the same value as the loudspeaker impedance (typically 4, 8, or 16 Ω) capable of dissipating the full audio output power (typically 10 or 15 W, but possibly higher).

2. Select the desired channel on the receiver being checked. The check can be performed on any channel, and normally needs checking on only one channel.

3. Set the receiver volume control to maximum and the receiver squelch control to the fully unsquelched position (fully counterclockwise on most communications receivers).

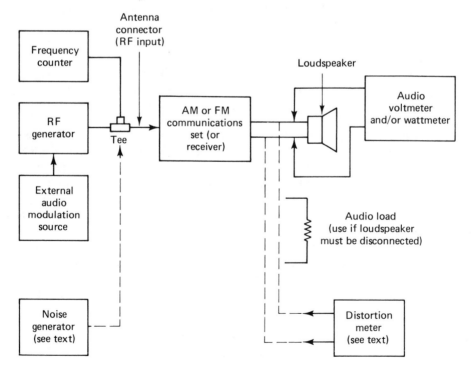

Figure 3-3   Test connections for AM and FM receiver checks.

4. If the receiver is equipped with adjustable RF gain, set the gain to maximum.

5. If the receiver is equipped with accessory circuits such as a noise limiter or ignition noise blanker, switch all such accessory circuits to off. This precaution may exclude some circuits as a possible source of trouble. Accessory circuit operation should be tested after all basic checks are performed.

6. With the receiver unsquelched and any noise limiter circuits off, check that there is receiver background noise on the loudspeaker. The total absence of background noise with the volume control at maximum indicates a totally dead receiver.

7. Set the RF generator to the approximate frequency of the receiver channel being checked, as close as the generator dial setting will quickly permit. You can use a frequency counter to set the RF generator, but it is not essential for this test.

8. Set the RF generator output level to 1000 μV. Turn on the RF generator internal modulation and set for 30% modulation. Some RF generators use 400 Hz internal modulation whereas others use 1000 Hz. Either frequency is satisfactory for an audio power check. If both frequencies are available, 1000 Hz is preferred.

9. Most RF generator dials are not calibrated sufficiently to permit adjustment to the exact frequency by dial setting alone. Even with a frequency counter, it is possible that the receiver is not exactly on frequency. In any event, the following procedure will permit you to set the RF generator to the exact receiver frequency.

Slowly adjust the fine frequency control of the RF generator back and forth about the correct frequency point as indicated on the frequency dial. When the correct frequency is approached, an audio tone will be heard from the speaker, and a reading will be obtained on the meter connected to the loudspeaker or audio load. The tuning range over which the tone and meter reading occurs may be very narrow, and the frequency tuning control of the RF generator may need very careful adjustment. Carefully adjust the frequency control for peak meter reading and peak audio output on the loudspeaker.

If the receiver is completely dead, no tone or meter reading will be obtained. Repeat the test on another channel, but with the RF generator output level set at maximum.

10. With the RF generator tuned for peak, at 1000 μV and 30% modulation, measure the power output on the meter connected to the loudspeaker or audio load. An audio wattmeter will provide a reading directly in watts. If an audio wattmeter is not available, use a voltmeter (VOM or digital, Sec. 2-3) and calculate the power using the equation

$$\text{power} = \frac{(\text{voltage})^2}{\text{loudspeaker impedance}}$$

For example, if you measure 4 V across a 4-Ω loudspeaker (or 4-Ω audio load), the power is 4 W. Figures 3–4 and 3–5 are included to help calculate power for typical loudspeaker impedances and voltages found in the audio sections of communications receivers. For example, if you find a 4-V reading on an 8-Ω loudspeaker, Fig. 3–4 shows that the power is 2 W.

    11. As a reference, a typical CB or mobile receiver will deliver about 2 or 3 W of audio power with the volume control set at maximum, and 1000 μV of RF at the input. Some base station communications receivers will deliver much more power. Always check the receiver specifications. Also, receiver audio power specifications usually include a maximum distortion figure (e.g., 2 W at less than 10% distortion). Distortion checks are described in Sec. 3–10.

    12. Once it has been established that the receiver is delivering adequate audio power, the following steps provide a quick check of audio-frequency response.

    13. Reduce the receiver volume setting to about one-half the rated maximum power.

    14. Note the reading on the meter connected to the loudspeaker or audio load. If the meter has a dB scale, set the volume control so that the reading is at some exact dB reading near the center scale of the meter.

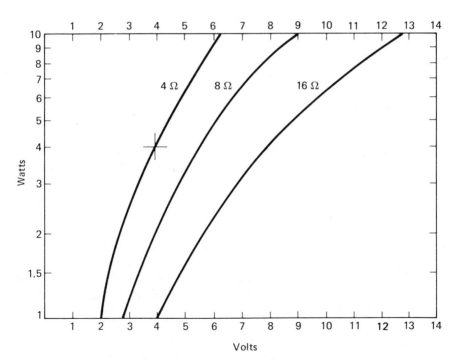

Figure 3-4 Audio voltage to audio power conversion chart, 1 to 10 W.

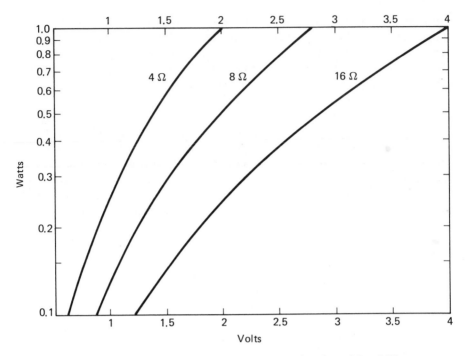

Figure 3-5   Audio voltage to audio power conversion chart, 0.1 to 1 W.

15. Leave the RF generator at the same RF frequency, but change the modulating frequency to 400 Hz, with 30% modulation. Note the reading on the meter in dB as compared to step 14.

16. Leave the RF generator at the same RF frequency, but change the modulating frequency to 2500 Hz, with 30% modulation. Note the reading on the meter in dB as compared to step 14.

17. The readings in steps 15 and 16 should be within the receiver manufacturer's frequency response specification, typically within +3 or −6 dB of step 14.

18. A more precise frequency response measurement may be made, if desired, by connecting a tunable audio generator to the external modulation input (if any) of the RF generator, selecting external modulation, and adjusting for 30% modulation with a 1000-Hz signal. The audio generator is then tuned from about 300 to 3000 Hz (or similar upper and lower limits specified by the receiver manufacturer), keeping the modulation of 30%, and noting the reading on the meter. Ideally, the frequency response should show no substantial dips or peaks, but a smooth response (within specifications) across the audio range (at least across the voice communications range).

3-5. RECEIVER SENSITIVITY CHECK

This check measures the weakest usable signal level at which the receiver will operate. The test is the best overall check of receiver performance that can be made, and should be performed immediately after the receiver audio power check (Sec. 3-4), since the equipment is connected and set up as required for the sensitivity check.

Receiver sensitivity is generally expressed as signal level required to produce a 10-dB signal-to-noise ratio (or to be more technically accurate, signal-plus-noise-to-noise ratio). For example, a typical receiver sensitivity specification is 1 μV for 10 dB (S + N)/N. This means that a 1-μV modulated signal into the receiver antenna input should produce an audio output at least 10 dB above the receiver noise level obtained with a 1-μV input signal without modulation. Many radio receiver specifications include the condition that the 10-dB (S + N)/N sensitivity to be obtained at some minimum audio output power. For example, a typical specification for overall receiver sensitivity is 1 μV for 10 dB (S + N)/N at 0.5 W audio output.

1. After performing the receiver audio power check, leave all connections and control settings as at the conclusion of that check. The RF generator should already be set to the receiver frequency with 1000-μV output and internal modulation of 30%.

2. Turn the receiver volume control to maximum and set the receiver squelch control fully unsquelched (fully counterclockwise).

3. Reduce the RF generator output level to a convenient low level such as 5 μV. If you think the receiver sensitivity may be normal, set the level to the manufacturer's specification (typically 1 μV or less). However, if receiver sensitivity is poor, it may be necessary to start with a higher value (such as 5 μV to get a reading in the following steps.

4. Reset the RF generator precisely on frequency. This is very important. If the RF generator is slightly off frequency, the test results will appear that the receiver has poor sensitivity. Rock the fine-tuning dial of the RF generator back and forth very slowly as the output level is reduced, and carefully adjust for peak reading on the audio output meter. Peak volume should also be heard from the loudspeaker (but always trust the meter reading rather than your ear). For most RF generators, the frequency must be repeaked after each change in RF output level control setting. The level control (usually an attenuator) tends to have some pulling effect on frequency.

If no meter reading is obtained with a 5-μV signal, receiver sensitivity is definitely poor (for modern communications receivers). Troubleshooting is required. One area of the receiver circuits to check first is the *frequency synthesizer*. Many modern communications receivers use a frequency synthesizer to produce the local oscillator signal. If the synthesizer signal is even slightly off frequency, sensitivity will appear to be very poor.

5. Note the audio meter reading on a convenient dB scale.

6. Switch the RF generator from internal modulation to no modulation (unmodulated carrier, CW, or whatever term is used for the RF generator modulation control setting).

7. The meter reading will drop. If the receiver has normal sensitivity, and a 5-μV signal is used, the meter reading should drop more than 10 dB from the step 5 reading. If you are not familiar with the dB scales of your particular meter, consider the following notes.

The dB scale of a meter is used to provide a convenient comparison between two readings, such as the 10-dB signal-to-noise ratio measurement. If both readings are taken on the same meter range, readings can be taken directly from the scale. For example, if the step 5 reading is $+2$ dB and the step 7 reading is -8 dB with both readings taken on the same range, the difference is 10 dB. However, for low meter readings (below about -5 dB with the modulation on) a more sensitive meter range should be used to get more accurate readings.

One convenient way to check for a 10–dB difference between steps 5 and 7 is to adjust the step 5 meter reading to 0 dB using the receiver volume control, regardless of the value in watts. The desired step 7 reading is then -10 dB (when the modulation is removed). However, this is not always possible since some receiver sensitivity specifications require that the test be made with a given audio output 0.5 W, 0.75 W, etc.), or with a given volume control setting.

8. Return the RF generator to internal modulation operation. Repeat steps 3 through 7 at progressively lower RF generator output levels until there is a 10-dB difference in meter readings between steps 5 and 7.

9. Note the setting of the attenuator (or output level control) on the RF generator. This setting, in microvolts, is the receiver sensitivity for 10 dB (S + N)/N, and should be equal to or lower than the receiver manufacturer's specification. For example, the RF generator output should be 1 μV (or less) for a specification of 1 μV for 10 dB (S + N)/N at 0.5 W of audio.

10. When you finally get the 10-dB sensitivity reading, note the audio output in watts. Use Figs. 3–4 and 3–5 if necessary. The audio output should equal or exceed the manufacturer's specification. For example, the audio output should be 0.5 W (or more) for a specification of 1μV for 10 dB (S + N)/N at 0.5 W of audio.

11. The full receiver sensitivity check need not be performed on all channels, unless you are having a particular problem with one or more channels. However, proper operation on all channels can be checked quickly as follows:

Leave the RF generator set at the 10-dB sensitivity level of step 9. With 30% internal modulation, note the audio meter reading. Select each receiver channel in turn. Tune the RF generator to each channel frequency and fine tune for peak audio meter reading. You should get the same reading for all channels (theoretically). If you find one or more channels that shows a substantially different reading (particularly a low reading) you have a good starting point for troubleshooting.

12. In most cases, it is only necessary to know if the receiver meets or exceeds the manufacturer's specification for sensitivity. This can be done quickly as follows. Set the RF generator output at the specified sensitivity level with 30% modulation (for example, 1 μV with 30% modulation). Set the receiver volume control to a convenient level as observed on the meter (0.5 W, 0 dB, etc.). Remove the generator modulation and observe the meter reading. If the reading drops 10 dB, or more, the receiver sensitivity is equal to, or better than, the 1-μV specification.

3-6. ADJACENT-CHANNEL REJECTION CHECK

Rejection of adjacent-channel signals is very important to prevent strong signals on adjacent channels from causing interference. This check is comparable to a *receiver selectivity test,* and measures the ability of the receiver to reject adjacent channel signals. The check can be used for both AM and FM receivers. A similar test for SSB receivers is described in Sec. 3-17.

Typically, adjacent-channel rejection is the same for all channels, and need be checked for only one channel. However, certain component failures can cause low adjacent-channel rejection only on specific channels. The check should be repeated for each channel showing adjacent channel interference. Typically, communications receiver channels are separated by 10 kHz or possibly 20 kHz.

1. Perform the receiver sensitivity check (Sec. 3-5), and leave connections and controls as at the conclusion of that check.

2. Set the receiver to the desired channel.

3. Leave the RF generator set for 30% internal modulation.

4. Tune the RF generator to the receiver frequency.

5. Set the RF generator output level to the 10-dB (S + N)/N level, which is 1 μV or less. Use the lowest possible signal level. Note the level for reference.

6. Adjust the receiver volume control for a convenient reference level on the audio meter, such as 0.5 W. Use a relatively low volume with respect to maximum rated audio.

7. Switch the receiver to the adjacent higher channel, but leave the RF generator tuned to the reference channel selected in step 2.

8. Increase the RF generator output level until the audio meter reads the same as the reference voltage level selected in step 6. To be sure that the RF generator remains precisely on the reference frequency, temporarily switch the receiver back to the reference channel (step 2) after the RF output level is readjusted, and retune the RF generator, if necessary. After tuning the RF

generator, switch back to the adjacent higher channel. Do not change the receiver volume or other controls.

9. Read the RF generator output level and compare the reading with step 5. The difference between the readings, in dB, is the adjacent channel rejection figure. This figure should be at least 30 dB for any receiver. A high-quality communications receiver may measure 60 dB or more. This figure should exceed the manufacturer's selectivity specification, which is usually stated for 20-kHz bandwidth. If the adjacent-channel rejection measures in the vicinity of 100 dB, the test results are suspect. The receiver probably has poor sensitivity. Use the lowest possible reference level (step 5) to reduce the probability of receiver desensitization.

10. Switch the receiver to the adjacent lower channel.

11. Normally, the audio meter should read the same as in step 8, which indicates that lower adjacent channel rejection equals higher adjacent channel rejection. If the audio meter reading is different from step 8, readjust the RF generator output level until the audio meter reading equals step 8. Read the RF generator level and compare with the step 5 level. The difference between the readings in dB is the lower adjacent channel rejection figure.

3-7. SQUELCH THRESHOLD SENSITIVITY CHECK

This test requires an extremely stable signal generator, preferably crystal-controlled. The check measures the weakest signal required to unsquelch the receiver when the squelch control is set at threshold. Perform this check immediately after the adjacent-channel rejection check (Sec. 3-6). Generally, the test need be performed only on one channel, or not more than three channels (low, middle, and high channels).

Squelch threshold sensitivity is measured in microvolts. The measured reading in microvolts should be equal to or less than the manufacturer's specification. Typically, squelch threshold sensitivity should be less than the specified 10-dB $(S + N)/N$ sensitivity of the receiver (which is typically less than 1 μV).

1. Set the receiver squelch control to minimum.

2. Set the signal generator exactly on frequency with 30% modulation at 1000 Hz.

3. Set the RF generator output to minimum.

4. If audio output corresponding to the generator modulation is observed at minimum RF generator output, switch the receiver channel selector to an adjacent channel so that only receiver noise is observed.

5. Set the receiver volume control to a convenient output level as observed on the meter or at the loudspeaker.

6. Adjust the receiver squelch control from minimum to threshold (the

point at which the receiver noise output just disappears). Typically, receiver noise should be reduced at least 20 dB when the receiver squelches.

7. Switch the receiver back to the test channel. If audio output corresponding to the signal generator modulation is observed, the squelch threshold level is less than the minimum generator output level.

8. If the receiver does not unsquelch, slowly increase the RF generator output level until it unsquelches and read the RF output level in microvolts. This is the squelch threshold sensitivity. Remember that any change in the RF generator output level may affect the frequency on many RF generators. Be sure that the RF generator is precisely on frequency. Repeat the check if there is any doubt that the most sensitive reading was obtained.

3-8. TIGHT SQUELCH SENSITIVITY CHECK

This test requires an extremely stable signal generator, preferably crystal-controlled. When the receiver is adjusted for tight squelch (squelch control fully clockwise), weak signals should be blocked, but strong, locally transmitted signals should pass. This check measures the signal strength required to un-squelch the receiver when the squelch control is set at tight squelch. Typically, the sensitivity should not exceed 1000 μV, but may be as low as 30 μV for some communications receivers. This check should be performed immediately after the squelch threshold sensitivity check (Sec. 3-7).

1. Leave all controls as at the conclusion of Sec. 3-7. The RF generator should already be set for internal modulation at 30% and should be exactly on the test channel frequency.

2. Set the RF generator output level to minimum.

3. Set the receiver squelch control to tight squelch (fully clockwise). The audio output meter reading should drop to zero.

4. Slowly increase the RF generator output level until the receiver un-squelches, at which time there will be audio output from the speaker and on the audio meter.

5. In step 4, the receiver may unsquelch at an unacceptably high signal level because the RF generator is pulled slightly off frequency during the measurement. To make sure that the most sensitive reading is obtained, use the following technique.

Reduce the RF generator output level until the receiver squelches. Reduce the squelch control setting so that the receiver unsquelches. Set the RF generator precisely on frequency (get a peak reading on the audio meter). Return the squelch control to tight squelch and slowly increase the generator output until receiver audio output is observed. Read the tight squelch sensitivity in microvolts from the RF generator output-level indicator.

3-9. AGC CHECK

This check verifies proper operation of the receiver AGC (automatic gain control) circuit. As the input signal level is changed from about 50,000 μV to 1 μV, the audio output level should not change more than 30 dB (typically). This check can be performed after the squelch checks, and need be performed on only one channel.

    1. Leave the equipment connected as for the receiver sensitivity check.

    2. The RF generator should be on frequency and adjusted for 30% internal modulation at 1000 Hz.

    3. Set the RF generator output level to 50,000 μV and retune as necessary so that the RF output is precisely on frequency (peak reading on the audio meter).

    4. If the receiver has an RF gain control, set the control to maximum.

    5. Set the receiver volume control for a convenient reference reading on the audio meter (0 dB, +10 dB, etc.).

    6. Slowly reduce the RF generator output level to 1 μV, noting that the audio meter reading drops smoothly (without sharp dips or peaks). The meter reading at 1 μV should not be more than 30 dB below the reference reading at 50,000 μV. As the meter reading approaches about -20 dB of the reference reading, retune the RF generator to make sure that it is on frequency.

3-10. DISTORTION CHECK

This check measures the percentage of audio distortion of a 1000-Hz test signal. The distortion specification for most communications receivers is rated in percent at a given audio output level (for example, less than 10% at 2 W). Distortion can be accurately measured only if the modulating signal is undistorted. Thus, if the RF generator modulation (internal or external) is distorted, it may appear that the receiver has more distortion than is actually the case. Before performing this distortion check, measure the RF generator modulation distortion using a distortion meter (Sec. 2-12.3). If the distortion is severe (something close to the desired maximum distortion figure for the receiver) use another modulating source with low distortion. Subtract any distortion in the modulating source from the measured receiver distortion. This check can be performed after the squelch checks, and need be performed only on one channel.

    1. Leave the equipment connected as for the receiver sensitivity check. Connect a distortion meter in parallel with the audio meter.

    2. Set the RF generator controls for 1000 Hz, 30% modulation.

    3. Set the RF generator output level to 1000 μV.

    4. Adjust the receiver volume control for the rated audio power output, as read on the audio meter. Use Figs. 3-4 and 3-5 if necessary.

5. Set up the distortion meter to measure 1000 Hz distortion, and read the percentage of distortion. The distortion should be less than the specified value, with the audio output at the rated wattage.

6. Readjust the receiver volume control to get maximum rated distortion (for the receiver being tested), as read on the distortion meter.

7. Read the audio output power on the audio meter. The audio power output (in watts) should equal or exceed the receiver specification at the maximum rated distortion.

3-11. RECEIVER ANL EFFECTIVENESS CHECK

Most communications receivers have an automatic noise limiter (ANL) circuit. The effectiveness of these circuits can be checked only in the presence of noise. Therefore, the RF generator used for an ANL check must also include a noise generator or signal. The effectiveness of an ANL circuit is measured by making a 10-dB (S + N)/N sensitivity measurement without noise (as described in Sec. 3–5), and again in the presence of noise. Sensitivity is degraded in the presence of noise. The effectiveness of the ANL circuit determines the degree to which sensitivity is degraded (typically less than about 10 dB for a good-quality communications receiver).

1. Leave the equipment connected as for the receiver sensitivity check (Sec. 3–5).

2. Set the RF generator controls for 1000 Hz, 30% modulation.

3. Set the RF generator output to the 10-dB (S + N)/N sensitivity level, as determined by the procedures of Sec. 3–5 (typically about 1 μV).

4. Turn on the noise generator. The audio meter reading should drop.

5. If the receiver is provided with an ANL switch, turn the ANL circuits on. The audio meter reading should increase when the ANL circuits are on.

6. Readjust the RF generator output level as necessary to get the 10-dB (S + N)/N reading (same audio meter reading as obtained in step 3). Generally, the RF output level must be increased to get the same audio meter reading in the presence of noise.

7. Compare the RF output level settings of steps 3 and 6. The sensitivity in the presence of noise should be within 10 dB of the sensitivity without noise. For example, if 1 μV is required in step 3 (no noise) to get the desired audio meter reading, no more than about 3 μV should be required in step 6 (noise) to get the same audio meter reading (since 10 dB represents a voltage ratio of about 3).

3-12. RECEIVER NOISE BLANKER EFFECTIVENESS CHECK

Many communications receivers have a noise blanker or noise eliminator circuit which is intended to eliminate most of the noise caused by electrical interference, such as poorly suppressed vehicle ignitions, electric motors, neon signs, and so on. The effectiveness of these circuits can be checked only in the presence of noise. Therefore, the RF generator used for a noise blanker check must also include a noise generator or source. The effectiveness of a noise blanker is measured by making a 10-dB (S + N)/N sensitivity measurement without noise (as described in Sec. 3–5), and again in the presence of noise. Sensitivity is degraded in the presence of noise. The effectiveness of the noise blanker circuit determines the degree to which sensitivity is degraded (typically less than about 3 dB for a quality communications receiver).

1. Leave the equipment connected as for the receiver sensitivity check (Sec. 3–5).

2. Set the RF generator controls for 1000-Hz, 30% modulation.

3. Set the RF generator output to the 10-dB (S + N)/N sensitivity level, as described by the procedures of Sec. 3–5 (typically about 1 μV).

4. Turn on the noise generator. The audio meter reading should drop.

5. If the receiver is provided with a noise blanker switch, turn the noise blanker circuits on. The audio meter reading should increase when the noise blanker circuits are on.

6. Readjust the RF generator output level as necessary to get the 10-dB (S + N)/N reading (same audio meter reading as obtained in step 3). Generally, the RF output level must be increased to get the same audio meter reading in the presence of noise.

7. Compare the RF output settings of steps 3 and 6. The sensitivity in the presence of noise should be within 3 dB of the sensitivity without noise (for a typical noise blanker circuit). As an example, if 1 μV is required in step 3 (no noise) to get the desired audio meter reading, no more than about 1.5 μV should be required in step 6 (noise) to get the same audio meter reading (since 3 dB represents a voltage ratio of about 1.5).

3–13. SSB TRANSMITTER RF POWER CHECK

This is normally the first SSB transmitter check performed. The test measures transmitter peak envelope power (PEP) to determine if the power level is normal. The test is similar to that for AM and FM transmitters, except SSB transmitters must be modulated (since these transmitters produce no power when not modulated), and a peak-reading instrument must be used to measure power.

1. Connect the equipment as shown in Fig. 3–6. Make certain that the RF wattmeter (Sec. 2–7) is capable of measuring the maximum PEP output power of the transmitter. If an RF wattmeter is not available, use a dummy load and a peak-reading meter to measure power as described in Sec. 2–6.

2. Set the transmitter to the first channel (or lowest frequency) to be

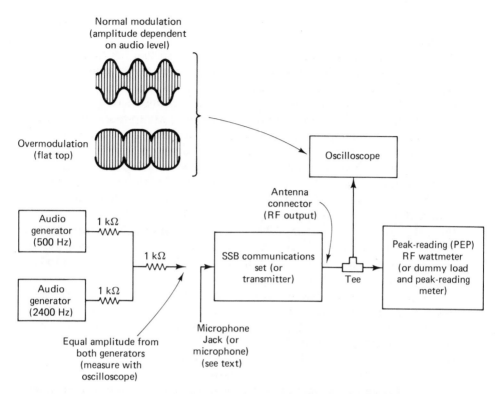

Figure 3-6 Test connections for SSB transmitter RF power (PEP) and modulation checks.

checked. Operate the transmitter controls to produce the maximum output power. In SSB operation, the carrier signal and one sideband are suppressed and all RF power is carried on one sideband. Thus, there is no RF output when the transmitter is unmodulated.

3. Apply two simultaneous, equal-amplitude audio signals for modulation, such as 500 Hz and 2400 Hz. The audio signals must be free from distortion, noise, and transients. The two audio signals must not have a direct harmonic relationship, such as 500 Hz and 1500 Hz. There are two alternatives for connecting the audio modulation signal sources to the transmitter. You can connect the modulation sources directly to the transmitter at the microphone input (or to a modulation input jack if the transmitter is so equipped). This provides the most stable and uniform modulation source. However, direct connection does not provide a check of the microphone. Also, on some transmitters, the set must be keyed by a push-to-talk switch on the microphone. The modulation audio generators can be connected to a loudspeaker. In turn, the microphone is placed directly over the loudspeaker. Use whichever method is most convenient.

4. If the transmitter is equipped with adjustable microphone gain, set the gain to midposition.

5. Secure the microphone over the speaker so that the speaker output drives the microphone with a constant tone. Adjust the modulation audio generators to frequencies of 500 and 2400 Hz, equal-amplitude, unless otherwise specified by the manufacturer's transmitter test instructions.

6. Key the transmitter with the microphone push-to-talk switch. Adjust the oscilloscope for a stable display of the modulation envelope. Note that if the vertical channel response of the oscilloscope is capable of handling the output frequency of the SSB transmitter (including the sidebands), the output can be applied through the oscilloscope vertical amplifier. If the oscilloscope is not capable of passing the SSB transmitter frequency, the transmitter output can be applied directly to the vertical deflection plates of the oscilloscope cathode-ray tube. However, as in the case of AM modulation test (Sec. 2–2), there are two drawbacks to this approach for SSB. First, the vertical plates may not be readily accessible. Next, the voltage output of the final RF amplifier may not produce sufficient deflection of the oscilloscope trace.

7. Check the modulation envelope patterns against the patterns of Fig. 3–6, or against patterns shown in the SSB service literature. Note that the typical SSB modulation envelope resembles the 100% AM modulation envelope, except that the amplitude of the entire SSB waveform varies with the strength of the audio signal. Thus, the percentage of modulation calculations that apply to AM cannot be applied to SSB.

8. Increase the amplitude of both audio modulation signals, making certain to maintain both signals at equal amplitudes. When peak SSB power output is reached, the modulation envelope will "flat-top" as shown in Fig. 3–6. That is, the instantaneous RF peaks of the SSB reach saturation, even with less than peak audio signal applied. This overmodulated condition results in distortion.

9. With the transmitter at maximum or peak power, but before overmodulation or "flat-topping" occurs, read the power in watts on the peak-reading RF wattmeter (Sec. 2–7) or on the dummy load and peak-reading meter (Sec. 2–6). This is the transmitter peak envelope power (PEP). Power output should meet the transmitter manufacturer's specifications, and must not exceed any applicable FCC limit for the unit being checked. For example, the peak RF output for class D CB transceivers in the SSB mode should not exceed 12 W.

10. If the SSB transmitter is capable of both lower sideband (LSB) and upper sideband (USB) operation, as is the case with many CB SSB units, check both the upper and lower sidebands of at least one channel, or all channels if desired. The PEP readings should be the same for all channels, on both upper and lower sidebands.

3-14. SSB TRANSMITTER MODULATION CHECK

This check of modulation quality displays the SSB transmitter modulation envelope for examination. For ease of understanding, the check is described as a separate test. However, in actual practice, the steps of the modulation check are often performed as part of the SSB transmitter RF power check (Sec. 3-13). Since the transmitter must be modulated to generate RF output, both can be checked simultaneously.

1. Perform steps 1 through 10 of the SSB transmitter RF power check as described in Sec. 3-13.

2. Reduce the modulation inputs to zero (set the gain controls of both audio generators to zero). The RF wattmeter (or dummy load and peak-reading meter) connected to the transmitter output should indicate zero.

3. Slowly increase the amplitude of both audio modulation signals, making certain to maintain both signals at equal amplitudes. When maximum power is approached, the amplitude of the waveform ceases to increase and the peaks flatten out. This is the overmodulation condition of Fig. 3-6, called "flat-topping." There should be a smooth transition from no output to full or peak output, with no sudden jerks or dips.

4. If the set is equipped with adjustable microphone gain, decrease the amplitude of both audio modulation signals to a point that produces one-half of the maximum measured RF power. Vary the microphone gain control from minimum to maximum, and again note that there is a smooth transition from minimum power to full power output.

5. The preceding tests are a convenient method for quickly verifying that the SSB performance is acceptable. If a detailed evaluation of SSB performance is required, such as that performed with a spectrum analyzer (Sec. 2-11), the two-tone signals from the audio generators should be applied to a microphone jack, rather than through a loudspeaker and microphone. The push-to-talk leads of the microphone jack can then be connected to a switch for keying the transmitter.

3-15. SSB TRANSMITTER FREQUENCY CHECK

This test measures the transmitter operating frequency in the SSB mode, and should be performed immediately after the SSB transmitter RF power and modulation checks. Since the RF carrier is suppressed in SSB operation, the frequency of the sideband signal is measured. If a 1000-Hz modulating signal is applied, the frequency of the upper sideband signal should equal the assigned carrier signal, plus 1000 Hz. A lower sideband signal should equal the assigned carrier signal, minus 1000 Hz. A stable single-frequency tone must be used for modulation during this check.

Most SSB communications sets are equipped with a *speech clarifier* adjustment, which is a fine frequency adjustment of the oscillator for clearest reception of SSB signals. Operation of the speech clarifier (sometimes called a *voice lock*) circuit is checked during this test. Although the speech clarifier circuit is used only while receiving, the circuit adjusts the oscillator (or frequency synthesizer), which is common to both the receiver and transmitter. It is easier to check operation of the speech clarifier circuit while measuring transmitter frequency.

1. Connect the equipment as shown in Fig. 3–7. As in the case of SSB modulation checks, you can connect the modulation source (audio generator) to the transmitter at the microphone input, or through a loudspeaker and the microphone. The direct connection gives you the most stable modulating signal source, but the loudspeaker–microphone method requires no special wiring and provides a simultaneous check of the microphone.

2. If the transmitter is equipped with adjustable microphone gain, set the gain to the midposition. Also set any speech clarifier control to the midposition.

3. Select the desired channel on the transmitter. Select the upper sideband, or USB mode, on sets where both USB and LSB are available.

4. Adjust the modulation source (audio generator) to 1000 Hz. Use the frequency counter to monitor the audio generator output to the transmitter. The audio generator output should be at 1000 Hz, within ±100 Hz. An accuracy of ±10 Hz is preferable.

5. Key the transmitter with the push-to-talk switch. Adjust the audio

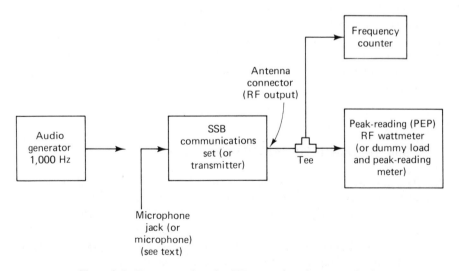

Figure 3–7 Test connections for SSB transmitter frequency checks.

generator output control until the RF output (as measured on the RF wattmeter or dummy load and meter) is equal to one-half the maximum rated PEP (as measured in Sec. 3–13). However, you will usually not get accurate frequency measurements unless the PEP is at least 1 W.

6. With the frequency counter connected to the RF output of the transmitter, read the frequency of the RF output signal. The indicated frequency should be equal to the assigned channel frequency, plus the modulating frequency of 1 kHz. For example, if you are monitoring class D CB channel 9, the assigned carrier frequency is 27,065,000 Hz, and the USB frequency with 1000-Hz modulation should be 27,066,000 Hz. If you are using a frequency counter with a five-place readout (which is typical for CB service), the USB reading for channel 9 should be 27.066 MHz.

7. Check the USB and LSB frequencies for each channel.

8. If the set is equipped with a speech clarifier or voice lock adjustment, it should be possible to adjust the frequency displayed on the frequency counter during steps 6 and 7. The typical range of speech clarifier adjustment should not exceed about ±1000 Hz from the assigned channel frequency.

9. To determine the adjustment range of the speech clarifier, get the USB frequency reading as outlined in steps 1 through 6 (with the speech clarifier control centered). Set the speech clarifier adjustment at the maximum counterclockwise position, and not the frequency reading observed on the frequency counter. Set the speech clarifier control to the maximum clockwise position, and note the frequency. The difference in readings is the total adjustment range of the speech clarifier.

3-16. SSB RECEIVER SENSITIVITY CHECK

This check measures the weakest usable signal level at which the receiver will receive SSB signals, and should be performed after the AM code checks (on receivers capable of both AM and SSB operation). As in the case of AM, SSB receiver sensitivity is expressed in microvolts for a 10-dB (S + N)/N ratio at a minimum audio level. For example, $0.5\text{-}\mu V$ for 10 dB (S + N)/N at 0.5 W of audio, which means that a $0.5\text{-}\mu V$ signal into the receiver antenna input should produce an audio voltage output at least 10 dB above the noise level with an audio power output of at least 0.5 W.

For SSB checks, an unmodulated carrier signal is injected from the RF generator at the sideband frequency (slightly above the assigned channel carrier frequency for USB, and slightly below the assigned channel carrier frequency for LSB). When the unmodulated signal from the RF generator mixes or "beats" with the reinjected carrier in the receiver, an audio tone is produced in the receiver output.

1. Connect the equipment as shown in Fig. 3–8.

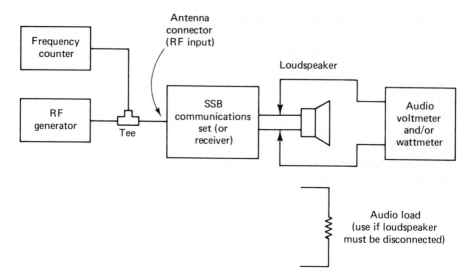

Figure 3-8   Test connections for SSB receiver checks.

2. Select the desired channel on the receiver. Select the upper sideband, or USB mode, on sets where both USB and LSB are available.

3. Set the receiver volume to maximum. If the receiver is provided with adjustable RF gain, adjust the gain for maximum.

4. Set the receiver squelch control fully unsquelched (fully counterclockwise). If the receiver is provided with accessory modes such as an automatic noise limiter or ignition noise blanker, turn all such modes off.

5. Set the RF generator to the unmodulated carrier mode. Adjust the RF generator output level to the 10-dB $(S + N)/N$ level. Start with the receiver sensitivity specification for the SSB modes (typically 0.5 μV or less).

6. Tune the RF generator for peak reading on the audio meter and peak volume on the receiver loudspeaker. The frequency of the audio output will vary as the RF generator frequency is changed. If no output is obtained on the meter and speaker, receiver sensitivity is poor, and a much greater RF generator level may be required.

7. Read the audio output level, in watts, from the audio meter. If an audio wattmeter is not available, use a voltmeter and calculate the power as described in Sec. 3-4, step 10. Adjust the receiver volume control for one-half of the receiver's rated maximum audio power. For example, if the receiver is rated at 2 W of audio, adjust for 1 W. If the audio output is less than one-half of the rated maximum, leave the volume control at maximum.

8. Note the audio meter reading on the dB scale.

9. Turn off the RF generator.

10. Again note the audio meter reading on the dB scale.

11. If step 10 is more than 10 dB below step 8, decrease the level of the RF

generator output and repeat steps 6 through 10. If step 6 is less than 10 dB below step 8, increase the level of the RF generator output and repeat steps 6 through 10. Continue adjustment until there is a 10-dB difference in meter readings between steps 8 and 10.

12. Note the setting of the attenuator (or output level control) on the RF generator. This setting, in microvolts, is the receiver sensitivity for 10 dB (S + N)/N.

13. It is not always necessary to measure the sensitivity in microvolts, but note whether or not the receiver meets the manufacturer's specification. In this case, set the RF generator to the specification level and note the audio meter reading. Then turn off the RF generator and again note the meter reading. If there is a 10-dB or greater difference between the meter readings, the receiver meets the specification.

14. After making the sensitivity reading, note the audio output in watts. Use Figs. 3–4 and 3–5 if necessary. The audio output should equal or exceed the manufacturer's specification.

15. Select the LSB mode and repeat the check. Sensitivity should be the same as for the USB mode.

16. Check USB and LSB mode sensitivity for each channel of operation. Sensitivity should be approximately the same for all channels.

3–17. SSB RECEIVER ADJACENT SIDEBAND REJECTION CHECK

This check measures the ability of an SSB receiver to suppress signals received on the opposite sideband. When the receiver is set for USB reception, any LSB input signals should be suppressed at least 40 dB. Similarly, USB signals should be suppressed at least 40 dB when the receiver is operating in the LSB mode. The check should be performed after, or during, the SSB receiver sensitivity check. The check can be performed on all, or only one channel, as desired.

1. Leave the connections as shown in Fig. 3–8. Repeat steps 1 through 6 of Sec. 3–17.

2. Note the audio meter reading for reference when the RF generator is tuned for a peak on the audio meter.

3. Switch the receiver to the LSB mode on the same channel.

4. Increase the output level of the RF generator until the audio meter reading is the same as in step 2, if possible.

5. It should require at least a 40-dB increase in RF generator output level to produce the same audio meter reading on the adjacent sideband. For example, if 1 μV is required in step 2, 100 μV (or more) should be required in step 4. If less than 40-dB suppression of the opposite sideband is measured, be sure the RF generator frequency is not shifted toward the opposite sideband by the output-level adjustment. Recheck RF generator tuning on the desired sideband.

6. Repeat the procedure, except adjust the RF generator for a peak on the LSB frequency, and measure USB suppression.

3-18. SSB RECEIVER SQUELCH SENSITIVITY CHECK

This check measures the minimum amount of on-frequency RF carrier required to unsquelch the receiver when adjusted at squelch threshold (the point that barely suppresses receiver noise), and at tight squelch (the point that requires a large signal to overcome). A typical SSB receiver requires 0.5 μV for the squelch threshold, and anything from about 30 to 500 μV for tight squelch. The receiver should not block strong signals, even when set at tight squelch.

The check should be performed after the sensitivity and sideband rejection checks, and is very similar to the squelch sensitivity checks for AM receivers, except that the receiver is operated in the USB or LSB modes, and an unmodulated signal is injected from the RF generator. The check can be performed on all, or only one channel, as desired.

1. Leave the connections as shown in Fig. 3–8.

2. Select the USB mode of operation on the desired channel.

3. Set the RF generator to the unmodulated carrier mode. Tune the RF generator frequency for maximum reading on the audio meter. Set the RF generator output level to the 10-dB (S + N)/N sensitivity level as determined in Sec. 3–16 (typically 0.5 to 1 μV).

4. Turn off the RF generator. Adjust the receiver volume control as necessary so that noise is heard on the loudspeaker.

5. Adjust the receiver squelch control to the squelch threshold (to the point where the noise is just squelched).

6. Set the RF generator output control to minimum, and turn on the RF generator (unmodulated carrier).

7. Slowly increase the RF generator output level until the receiver unsquelches. There should be at least a 20-dB difference in the audio meter reading between the squelched and unsquelched condition. In the unsquelched condition, audio output should be at least one-tenth of the receiver's rated maximum audio output.

8. Note the setting of the attenuator (or output level control) on the RF generator. This setting, in microvolts, is the SSB squelch threshold setting of the receiver. The reading should be equal to or less than the receiver specification for SSB squelch threshold. Typically, this value is 0.5 μV or less.

9. Adjust the receiver squelch control for tight squelch (fully clockwise).

10. Increase the output level of the RF generator until the receiver unsquelches. To make sure that the RF generator remains on frequency when the output level is increased, temporarily reduce the squelch setting and retune the RF generator for peak reading on the audio meter, then return the squelch control to tight squelch.

11. Note the setting of the attenuator (or output level control) on the RF generator. This setting, in microvolts, is the tight squelch sensitivity. The reading should be equal to or less than the receiver specification, which is typically in the range 30 to 500 μV.

12. Switch the receiver to the LSB mode on the desired channel, and repeat the procedures of steps 1 through 11.

3-19. ANTENNA CHECK AND SWR MEASUREMENT

The antenna check is one of the most important performance checks that can be made when troubleshooting radio communications equipment. The quickest and most effective antenna check is an SWR measurement, since such measurements show just how well the communications set, antenna, and antenna cable are matched and tuned. The procedure for making SWR measurements is generally very simple, but depends on the type of SWR meter. Some communications receivers have built-in SWR meters. In most cases, SWR measurements are made using an external SWR meter and the basic operating procedures described in Sec. 2-9. We will not give further, more detailed SWR measurement procedures here since the exact procedures depend on the type of SWR meter. Instead, we include a series of notes that apply to all types of SWR measurements.

As described in Sec. 2-9, the basic SWR measurements are as follows. The SWR meter is connected between the communication set (transmitter output) and the antenna. The transmitter is keyed, and the SWR meter needle is aligned with some "set" or "calibrate" line (usually near the right-hand end of the meter scale). A switch is then set to the SWR measurement position, with the transmitter still keyed, and the SWR meter needle drops back to read the SWR.

A low SWR reading is desired since this indicates that the transmitter is operating at maximum effectiveness, and receiver performance is optimum. A low SWR results when the antenna is properly tuned to the operating frequency and there is a close match of impedance between the transmitter output, antenna cable, and antenna. Not all antennas are tunable. When the antenna is tunable, the SWR measurement can be used to find the optimum frequency. The SWR is measured and, with the transmitter still keyed, the antenna loading coil (or other tuning element) is adjusted for minimum SWR. However, always follow the antenna tuning procedure recommended in the service literature. The antenna is normally tuned for minimum SWR on the center operating frequency (for example, on channel 11 or 12 of a 1-to-23 channel CB). Compact mobile antennas and high-gain beam-type base station antennas are frequency sensitive, and show variations in SWR as the transmitter is keyed on all channels (particularly when there is a wide frequency range for the lowest- to the highest-frequency channel).

The SWR measurement must be made using the antenna and antenna cable that are normally used with the communications set. Be sure the SWR test includes all the antenna cable and connectors that are normally used. An SWR measurement is essential at the time of installation and should always be performed after repairs to the set are completed. The check is also needed if damage to the antenna or antenna cable is suspected. A periodic SWR measurement will detect any gradual deterioration and assure continued high performance. A damaged antenna or cable, corroded connectors, or a similar problem can cause a very high SWR. In turn, a high SWR often causes premature failure of final RF amplifier transistors in solid-state transmitters.

3–20. RECEIVER FREQUENCY RESPONSE CHECK

This check measures receiver audio-frequency response. An audio-frequency specification of 300 to 3000 Hz usually means that all audio frequencies from 300 to 3000 Hz at a given input level should produce audio outputs that are within 3 dB across the entire range. A 1000-Hz reference is often used to measure the 3-dB level. However, the point of reference can be the frequency within the response band at which the maximum output level is developed.

The check is performed by applying a constant-amplitude modulated test signal to the receiver input. The modulation percentage is maintained at a constant value, and the receiver audio output level is observed as the modulation frequency is varied.

1. Connect the equipment as described in Sec. 3–4. Use the audio generator connected to the external modulation input of the RF generator, as described in step 18 of Sec. 3–4.

2. Select the desired channel on the receiver. The check can be performed on any channel. Unsquelch the receiver. Adjust the receiver volume and RF gain (if adjustable) to the midposition. On AM/SSB/FM receivers, use the AM mode. Turn off all accessory mode switches (noise blankers, etc.) on the receiver.

3. Set the RF generator for external modulation. Set the audio generator to 1000 Hz. Adjust the audio signal generator level and the modulation adjustment of the RF generator for 30% modulation. Adjust the RF generator output level for 1000 μV.

4. Adjust the receiver volume control for a convenient reference level on the audio meter (try 0 dB). Use a volume well below the maximum capability of the receiver to minimize distortion.

5. Tune the audio generator across the band of audio frequencies specified in the receiver service literature. In the absence of any specifications, tune across the audio range from 300 to 3000 Hz. Readjust modulation as required to maintain 30% modulation as the frequency of the audio generator is changed.

6. Read the audio meter as the frequency of the audio generator is changed. The meter reading should not change more than 3 dB over the entire range from 300 to 3000 Hz.

3-21. PUBLIC ADDRESS MODE CHECK

Some communications sets (particularly CB sets) are equipped with a public address (PA) mode. In this mode, the microphone audio is amplified and applied to a separate PA loudspeaker. This check confirms proper operation of the PA mode.

1. Connect the equipment as shown in Fig. 3-9. Operate the communications set in the PA mode. Make certain that there is no feedback from the PA loudspeaker to the microphone. If necessary, disconnect the PA loudspeaker and replace it with a dummy load of suitable impedance (typically 4, 8, or 16 Ω).

2. Adjust the audio generator to 1 kHz. Increase the audio output to a level where the 1-kHz tone can be heard on the test loudspeaker.

3. Place the microphone over the test loudspeaker, and press the push-to-talk-button.

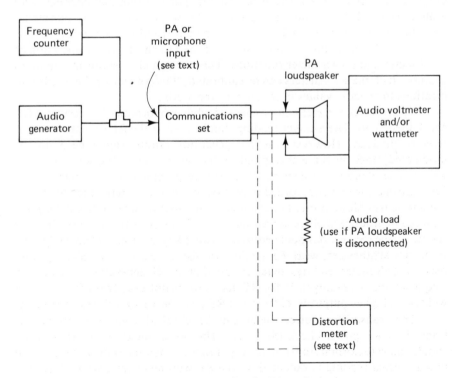

Figure 3-9 Test connections for PA mode checks.

4. If the microphone gain is adjustable, vary the gain and note that the audio power reading on the audio meter varies. If the microphone gain is not adjustable, vary the audio generator output amplitude and note that the audio power reading varies.

5. Measure the audio power as indicated by the audio meter. If the audio meter does not read out in watts, measure the voltage and calculate the wattage (as described in Sec. 3–4, step 10). Compare the audio power reading against that specified for the PA mode in the service literature.

6. If desired, the distortion can be measured while operating in the PA mode. Connect the distortion meter to the PA loudspeaker or dummy load. Set up the distortion meter to measure 1000-Hz distortion, and read the percentage of distortion. The distortion should be less than the specified value, with the audio output at the rated wattage.

3-22. S-METER/POWER METER CHECK

Many communications sets are equipped with an S-meter that indicates received carrier signal strength, and a power meter that indicates transmitter RF output power. In most sets, the meter is a dual-purpose unit that operates as an S-meter while receiving, and as a power meter while transmitting. Normally, these meters give relative indications rather than specific values in microvolts and watts. The procedures in this test can be used to check proper operation of the S-meter and power meter functions. This test can also establish a standard against which the set's meters can be calibrated, thus converting the relative indications to specific values in microvolts and watts.

The receiver S-meter can be checked while performing the receiver AGC check (Sec. 3–9). The S-meter reading should vary as the RF generator output level is changed. However, the RF generator output signal need not be modulated, since the S-meter responds to the carrier signal. If desired, note the RF signal level required (in microvolts) for each increment on the S-meter scale. This information can be recorded or plotted on a graph, and retained for future use during troubleshooting. For example, if an input between 100 and 200 μV is required to produce an S9 reading, and you find a receiver that requires 1000 μV for an S9, the receiver has poor sensitivity (caused by improper adjustment, off-frequency synthesizer, weak RF section transistors, etc.). When making comparisons of S-meter readings, make certain that the RF generator is precisely on frequency (peak readings). If the RF generator is not exactly on frequency, it will take a higher output level from the RF generator to get the same reading.

The transmitter power meter can be checked while performing the transmitter RF power check (Sec. 3–1). The power meter of the transmitter should indicate normal transmitter output power when the transmitter is keyed and a normal reading is obtained on the RF wattmeter (or dummy load and meter). If desired, note the power level (in watts) on the RF wattmeter for each increment on the transmitter power-level meter. This information can be recorded or plotted on a graph, and retained for future use during troubleshooting.

If the communications set operates on both AM and SSB modes, the power meter will probably have a different range or scale for each mode. After the power meter has been checked in the AM mode, check the SSB scales while performing the SSB transmitter RF power check (Sec. 3–13) and the SSB transmitter modulation check (Sec. 3–14). Note that the power meter and the RF wattmeter indications vary as the modulation varies. Note the power (in watts) for each increment of the power meter.

3-23. EFFECTS OF VOLTAGE CHECK

All the checks described thus far in this chapter (Secs. 3–1 through 3–22) should be performed with the communications set operated at the rated input voltage. For example, a typical mobile CB set requires an input or power supply voltage of 13.8 V (direct-current). It is good troubleshooting practice to note the effects on test results when the power-supply voltage is varied, particularly the effects of low voltage. Where possible, repeat all the tests and vary the power-supply voltage over the range 11 to 15 V. Note the test results at various voltage levels. However, make certain not to exceed the maximum rated power-supply input voltage (typically about 15 V for a mobile communications set).

Power-supply voltage changes should have very little effect on transmitter frequency. However, transmitter power, receiver sensitivity, and audio or PA power may be affected significantly, even by a 1-V change in the power supply. Testing at low and high voltages can occasionally reveal a fault that is undetected at normal voltage levels.

3-24. COMMUNICATIONS SET TROUBLESHOOTING APPROACH

The remainder of this chapter is devoted to some practical tips on using the check procedures described thus far to isolate troubles in communications sets. Of course, the most valuable troubleshooting tool is a strong knowledge of circuit fundamentals and functional operation of the set. Proper use of test equipment can convert that knowledge into faster troubleshooting for technicians of all experience levels. The techniques presented here demonstrate the guideline for developing a logical, systematic approach to troubleshooting. The procedures are presented in a logical sequence based on an analysis of the set's operation.

Although each manufacturer (and, in fact, each model) has its own design variations, there is a basic similarity among most communications sets. For example, Fig. 3-10 shows a block diagram of a typical AM-only CB transceiver, while Fig. 3-11 is the block diagram of a typical AM/SSB transceiver. These transceivers, which are typical of present-day communications sets, use a dual-conversion receiver with 7.8-MHz and 455-kHz IF circuits, synthesizer-type transmit/receive oscillator, and an audio circuit that is common to transmit, receive, and PA modes of operation. The SSB receiver IF circuits and transmitter RF circuits are independent of the AM circuits, but several SSB circuits are common to transmit and receive operations. These are the most common basic designs found in present-day CB transceivers.

The troubleshooting procedure described here is tailored to the circuit designs shown in Figs. 3-10 and 3-11 in order to demonstrate more clearly the analyzing technique. However, the basic technique can be modified to troubleshoot communications sets of almost any design. As you become more experienced and proficient, you are encouraged to develop your own technique and shortcuts which can further reduce servicing time and improve profitability. As discussed in Chapter 1, by carefully observing the symptoms, it is often possible to go directly to a specific checkout procedure and bypass unrelated checks.

Virtually all CB transceivers (and most communications sets of any type) built in recent years are fully solid-state units that operate directly from a 12-V vehicle battery (or 120-V alternating current for most base stations). Therefore, no effort is made to include specific troubleshooting information for vacuum-tube circuits or dc-to-dc power supplies (converters). However, many of the checkout procedures are applicable to vacuum-tube equipment as well as solid-state circuits.

The troubleshooting procedures here isolate the defect to a small area consisting of only a few parts. Conventional voltage and resistance measurements should then be made within the suspected circuit to locate the defective part. Refer to the service manual or the set being serviced for data such as a schematic diagram, normal RF and d-c voltages, and test specifications. As discussed in Chapter 1, a detailed block diagram of the set is a valuable tool for rapid trouble isolation.

3-25. MOBILE COMMUNICATIONS SET INITIAL CHECKS

When troubleshooting mobile communications sets, it is advisable to perform a few checks to eliminate all items external to the set before removing the set from the vehicle for bench servicing. External items such as the power cable, antenna, antenna cable, microphone, and external speaker are often more subject to physical damage than the set, and must always be considered as a likely source of trouble. The following checks will quickly verify if external items are good or bad.

If the set is already removed from the vehicle and is in the service shop for

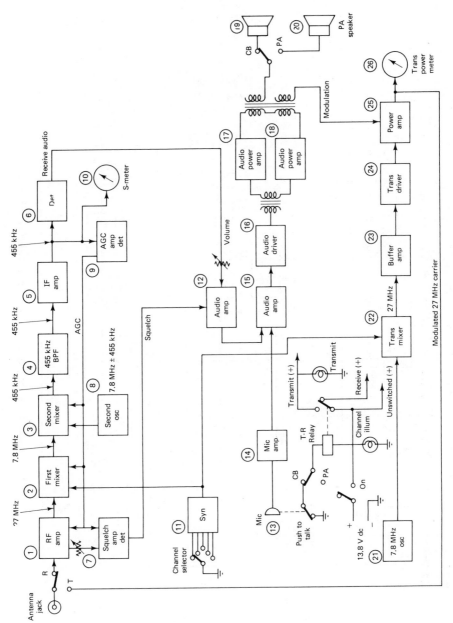

Figure 3-10 Block diagram of typical AM-only CB transceiver.

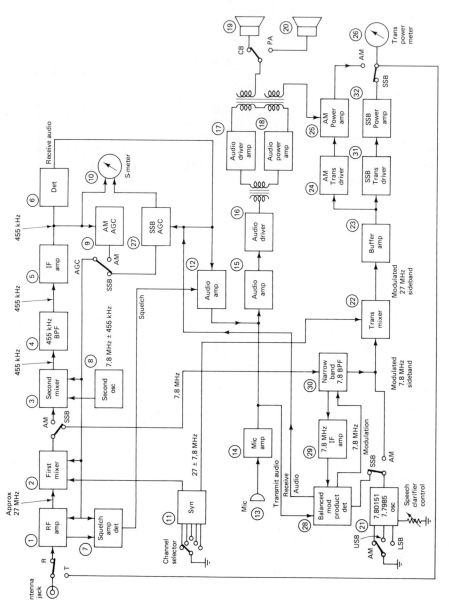

Figure 3–11 Block diagram of typical AM/SSB CB transceiver.

checkout, proceed with the bench check. However, if the set indicates normal operation on the bench, these in-vehicle checks may be necessary when the set is returned to the vehicle. For AM/SSB sets, perform all of the following checks in the AM mode.

3–25.1 Power Cable Check

1. Turn on the set. Be sure that the vehicle ignition switch is also on, if required for radio operation.

2. On most sets, there is some type of indicator light which turns on when power is applied. Sometimes the indicator light is in the form of channel selector illumination or meter illumination. Note whether the indicator light is on.

If the indicator light is on, the power cable is probably okay. If the indicator light does not come on when power is applied, check the fuse.

If the fuse is good, power apparently is not getting through the power cable to the set. Make voltage and continuity checks on the power cable and the related connectors. If power is available to the set power connector, the problem is within the set.

If the fuse is open, replace it with a new fuse of the proper rating. If the new fuse blows, there is an apparent overload or short circuit (probably in the receiver section).

If the new fuse does not blow, key the transmitter. If this causes the fuse to blow, there is an apparent overload or short circuit (probably in the transmitter section).

If the new fuse does not blow when the transmitter is keyed, normal operation is restored, at least temporarily. Operate the set in both the receive and transmit condition while driving over a rough street. If the fuse does not blow, the likelihood of future fuse failure is greatly reduced. If the new fuse blows during the test drive, there is an apparent intermittent short within the set.

3–25.2 External Speaker Check

If the installation does not include an external speaker, skip to the antenna check (Sec. 3–25.3).

1. Unsquelch the receiver and turn up the volume. If receiver background noise or received signals are heard on the external speaker, it is operating.

2. If received signals are heard on the external speaker, but severe audio distortion is noted, as could be produced by a speaker with a torn cone, disconnect the external speaker and listen to the internal speaker (or connect a substitute speaker). Do not operate a receiver without any speaker (or without a dummy load of appropriate impedance), as that would probably cause the transistors in the final audio section to be destroyed. If no distortion is noted on the

internal or substitute speaker, replace the external speaker. If distortion is noted on both speakers, the trouble is within the set.

3. If nothing is heard from the external speaker, either the set or the speaker could be the cause. Disconnect the external speaker and listen from receiver noise or received signals on the internal speaker. (If nothing is heard, and you do not know if the set has an internal speaker, connect the substitute speaker to the external speaker jack. Use the PA speaker if the vehicle is so equipped.) If noise or received signals are heard on the internal or substitute speaker, the trouble is in the external speaker or its cable or connector. If nothing is heard on the internal or substitute speaker, it is unlikely that both speakers are defective. Take a wild guess that the trouble is within the set.

3-25.3 Antenna Check

This check cannot be completed unless there is transmitter power output. If any step shows the need for troubleshooting the set, the antenna check should be delayed until the set is returned to the vehicle.

1. If the set is equipped with a transmitter power meter, key the transmitter and observe the meter indication. (If the transmitter cannot be keyed, skip to the microphone check, Sec. 3-25.4.) If abnormal power output is indicated, you can generally omit the antenna check, because the problem is probably in the set. However, an extreme problem in the antenna (shorted cable, etc.) can cause an abnormal power indication.

2. If the transmitter power output reading is normal, or if the set is not provided with a transmitter power output meter, connect an external test SWR/power meter between the set and the antenna.

3. Key the transmitter and check forward power (or a power indication on the external meter when the meter is not in the SWR mode). If abnormal power output is indicated on the external meter, the trouble is probably in the set (or there is an extreme antenna problem).

4. Perform the antenna SWR measurement (Secs. 2-9 and 3-19).

If the SWR is less than 1.2, the antenna and antenna cable are excellent. If the SWR is from about 1.2 to 1.5, the antenna and cable are probably satisfactory.

Higher SWR readings indicate a mismatched condition which may be due to a damaged antenna, corroded or improperly fitted connectors, excessive moisture, or crushed antenna cables.

An extremely high SWR indicates an open-circuit or short-circuit condition, such as an antenna cable cut in two (or disconnected) or the inner and outer conductors of the antenna cable shorted together.

3-25.4 Microphone Check

A microphone check needs to be performed only if one or more of the following symptoms is indicated.

Transmitter cannot be keyed.

No transmitter modulation.

No output in the PA mode.

Disconnect the microphone and connect a known-good substitute microphone.

If no change in symptoms is noted, the trouble is within the set.

If the symptom is corrected when using the substitute microphone, the original microphone is defective.

If no substitute microphone is available, or the microphone is the wired-in type, remove both the set and the microphone from the vehicle for bench servicing.

3-25.5 PA Speaker Check

This check is applicable only if the installation includes a PA speaker, and the set includes a PA mode.

1. Operate the set in the PA mode. If the PA announcement is heard on the PA speaker (without excessive distortion), the PA speaker is ok.

2. If there is excessive distortion in the PA speaker, such as could be produced by a torn cone, disconnect the PA speaker and connect a substitute speaker. If there is no distortion in the substitute speaker, the PA speaker is defective. If there is distortion in both speakers, the trouble is in the set (and there will probably be distortion in the receiver audio and transmitter modulation).

3. If nothing is heard on the PA speaker, disconnect the PA speaker and connect a substitute speaker. If PA announcements can be heard on the substitute speaker, the PA speaker is defective. If PA announcements cannot be heard on either speaker, the trouble is probably in the set (but could be in the wiring between the set and PA speaker).

3-25.6 Conclusion of Initial Checks

If all items external to the communications set are good, the fault is in the set. Remove the set from the vehicle for bench servicing.

Whenever repairs have been completed, and the set is reinstalled in the vehicle, always perform an antenna SWR check. The installation should be adjusted for minimum SWR by tuning the transmitter to match the antenna (or vice versa, or both). Not all antennas are tunable. Always follow procedures recommended in the service manual for tuning the transmitter. If you cannot get an SWR below about 1.5 on all channels, and preferably better on the middle channels, look for problems in the antenna installation. A high SWR can cause burnout of the transmitter final RF amplifier, so if you service a set with such a condition, look for antenna problems.

If a bench check indicates normal performance, yet the set performs poorly in the vehicle and you have tuned for minimum SWR and maximum power output, try checking the following items:

1. Poor ground connection
2. Too much resistance in the power cable. Voltage into the set substantially below 13.8 V (or below the recommended power-supply voltage)
3. Vehicle battery voltage low or poor voltage regulation from the vehicle regulation system
4. Too much ignition interference and electrical disturbance, due to inadequate noise suppression on the vehicle
5. Antenna poorly located for good propagation characteristics

3-26. BASE STATION INITIAL CHECKS

The initial checks for a communications base station are essentially the same as for a mobile station. Often, base station equipment will have more meters for monitoring radio performance. Also, base stations usually operate from a 120-V a-c supply. In many cases, this power supply is converted to 13.8-V d-c, which is the same as the vehicle battery voltage for most mobile communications sets.

As with mobile sets, eliminate all items as possible troubles before starting a bench check on a base station.

1. Connect the base station to the power source.

2. Turn on the set. If the channel selector or meter illumination lamp is on, assume that the power-supply voltages are correct.

3. If the channel selector and/or meter illumination are not on, check the fuse and replace if necessary. If a new fuse restores operation, proceed to step 4. If a new fuse does not restore operation, remove the base station set, including the power supply, for bench servicing.

4. Key the transmitter. If the transmitter RF power output is normal, check the antenna SWR (with built-in or external SWR meter). Generally, base station SWR is better than for mobile stations.

5. Check the base station microphones by substitution, or include them in the bench check.

3-27. CURRENT DRAIN CHECKS

When a mobile communications set is brought to the service bench for checkout, the set is connected to a power supply. Most power supplies include output voltage and current meters. Typically, voltage should be adjusted to 13.8 V, and current limiting (a feature found on many power supplies) should be adjusted to the required current. If the set has been blowing fuses or indicates a high current drain, find the short circuit or overload and repair it before continuing with other checks. It is advisable to note the current drain for each set when the set is initially connected to the power supply. Refer to the service manual for maximum current drain for a specific set. For a typical CB set, maximum current drains are in the vicinity of:

Standby (receiver unsquelched)	500 mA
Receive (full-rated audio)	1.5 A
Transmit	2.2 A

3-28. DEFINING THE SYMPTOMS

Although proper test equipment can be used to perform a complete diagnosis of an ailing communications set without a description of the symptoms, a good description of the symptoms can often lead you directly to the problem. Servicing time is reduced by eliminating the time required for a complete checkout of all possible symptoms.

When a service order for a communications set is taken, get a full description of the symptoms if possible. This description should be defined as precisely as possible. Ask the owner or operator additional questions, if necessary, to refine the description. For example, a symptom of "poor reception" could include an entire range of symptoms from "weak audio" to "short range (poor receiver sensitivity)," "adjacent channel interference (poor receiver selectivity)," or "garbled voice (distortion)." One of these terms is much more precise in defining the malfunction. More important, the term could further isolate the problem area in which the trouble is located.

Unless the servicing job is to be started immediately, jot down a description of the symptoms in correct technical terms. Keep the note with the equipment for reference when the troubleshooting job begins.

The troubleshooting procedures in the remainder of this chapter are grouped by symptoms. The symptoms alone localize the trouble within a portion of the set. The troubleshooting technique that isolates the malfunction in the shortest time does not include checks in circuits that are not related to the symptom. On combination AM/SSB sets, the usual practice is to fully check the AM mode of operation completely and correct any malfunction before checking the SSB mode. Of course, this does not apply if the symptom is "failure in the SSB mode only."

3-29. SERVICE BENCH DIAGNOSTIC CHECK

The following general procedure is used to determine what is wrong with the set if no description of symptoms is available, or if you wish to verify the symptoms and check for additional symptoms. The procedure may also be used if you wish to perform a complete performance check of the set and correct any subnormal performance.

1. Connect the set into the basic test equipment configuration as shown in Fig. 3–1. Use the test connections shown in Figs. 3–1 through 3–11, as necessary, to perform the test procedures described in the following steps.

2. Connect an external loudspeaker in parallel with the set's internal loudspeaker.

3. Set the operating controls as follows: Select any channel, set the squelch control to the fully unsquelched position, set the RF gain control (if any) to maximum, set any accessory mode switches (such as noise limiter or automatic noise control) to off, and select the AM mode of operation.

4. Turn on the set, and turn up the volume until a strong receiver noise is heard on the loudspeakers. If noise is heard on the external loudspeaker but not on the set's internal speaker, you have already found one of your problems. If no receiver noise is present on either loudspeaker, adjust the volume control to about three-fourths of the maximum position.

5. Perform as many steps shown in Fig. 3–12 as is required to be directed to a troubleshooting procedure. Use the referenced troubleshooting procedure to isolate and correct any malfunction. For example, the first step shown in Fig. 3–12 is to "listen for receiver noise." If there is no receiver noise, the next step is to "perform transmitter RF power check, as described in Sec. 3–1." If the transmitter power check is normal, the next step is to refer to the troubleshooting procedure for "radio does not receive, as described in Sec. 3–30." If there is "no output on any channel" during the transmitter RF power check, the next step is to refer to the troubleshooting procedure for "radio does not transmit or receive, as described in Sec. 3–31." If there is "no output on some channels" during the transmitter RF power check, the next step is to refer to the troubleshooting procedure for "radio does not transmit or receive on some channels, as described in Sec. 3–32." Now assume that during the first step of Fig. 3–12 receiver noise is "low or normal," instead of "no receiver noise." Then the next step is to "perform receiver audio power check, as described in Sec. 3–4."

6. If troubleshooting and repair are required during the steps of Fig. 3–12, recheck operation of the set by repeating the steps of Fig. 3–12, starting at the beginning. Repeat as many times as required until the test results end with "check SSB mode of operation for AM/SSB sets."

7. If an AM/SSB set is being checked, perform as many steps of Fig. 3–13 as required to be directed to a troubleshooting procedure. Use the reference troubleshooting procedure to isolate and correct any malfunction (as you did during step 5 of this section). If an AM-only set is being checked, skip this step and proceed to step 9.

8. If troubleshooting and repair are required during the steps of Fig. 3–13, recheck SSB operation by repeating the steps of Fig. 3–13, starting at the

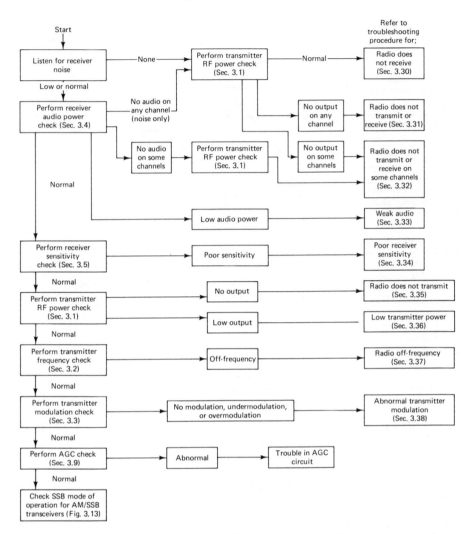

Figure 3-12 Service bench diagnostic check for AM mode.

beginning. Repeat as many times as required until the test results end with "SSB performance normal."

9. Perform as many steps of Fig. 3-14 as required to be directed to a troubleshooting procedure. Use the referenced troubleshooting procedure to isolate and correct any malfunction (as you did during step 5 of this section).

10. If any troubleshooting and repair are required in step 9, recheck the performance of the set by repeating the steps of Fig. 3-14, starting at the beginning. Repeat the steps as many times as required until the test results end with "all performance normal."

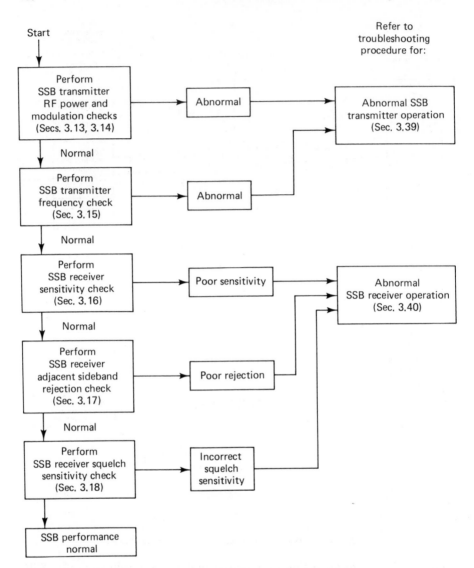

Figure 3-13   Service bench diagnostic check for SSB mode.

3-30. TROUBLESHOOTING PROCEDURE FOR "RADIO DOES NOT RECEIVE" SYMPTOM

Use this troubleshooting procedure when there is no received audio and the transmitter RF power is normal. More specifically, use this procedure if there is no receiver noise, or if an audio output cannot be obtained when a strong modulated carrier is applied to the receiver.

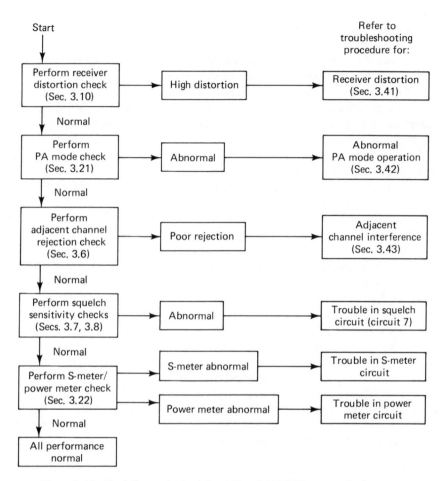

Figure 3-14 Final diagnostic check for AM and AM/SSB communications sets.

The symptom can be caused by failure in almost any portion of the receiver (circuits 1 through 9, 12, and 15 through 19 in Figs. 3-10 and 3-11).

It should be noted that if there is no receiver noise, it is probable that the fault is in the 455-kHz IF or audio sections of the receiver. If some receiver noise is present, the problem is more likely to be in the RF or first IF sections. An open circuit in the RF circuits often does not cause complete failure to receive. High-strength RF test signals can often be coupled through an open circuit with sufficient strength to produce an audio output. Of course, receiver sensitivity is poor. On the other hand, a short in the RF circuits can often give a symptom of no receiver audio.

The following steps isolate the problem to a much smaller area.

1. Perform the PA mode check (Sec. 3-21).

If the PA mode operates, the audio section is proven good (circuits 15 through 18). Proceed to step 2 if the set is not equipped with an S-meter, or to step 3 if the set has an S-meter.

If the PA mode does not operate, the audio section is the problem area (circuits 15 through 18), and can be further isolated as follows.

Operate the set in the PA mode. Connect an audio generator to a loudspeaker. Place the microphone over the loudspeaker. Set the audio generator to 1 kHz. Measure the audio signal using an oscilloscope at circuits 15, 16, 17, and 18 (in that order) until the loss of signal is noted. The circuit at which signal loss is noted is probably defective. Check the voltage, resistance, and components of that circuit.

2. If the set is not provided with an S-meter, inject a modulated 455-kHz IF signal into the detector (circuit 6).

If audio output is obtained, the detector (circuit 6) and the first IF (circuit 12) are probably good. Proceed to step 4.

If no audio output is obtained, measure the audio signal on an oscilloscope at each accessible measurement point in the signal path from the detector (circuit 6) through the first audio amplifier (circuit 12), and note the point of signal loss. If the input to circuit 12 is normal, but no output is measured, check the d-c voltages of the associated squelch switch circuit. Certain component failures can cause the circuit to remain in the fully squelched condition at all times.

3. If the set is provided with an S-meter, apply a 1000 μV, on-frequency carrier signal from an RF signal generator, and note the S-meter reading.

If a strong S-meter reading is obtained, the RF and IF sections are operating (circuits 1 through 5). The trouble is isolated to the detector (circuit 6) or first audio amplifier (circuit 12). Inject a modulated 455-kHz IF signal into the detector and troubleshoot as instructed in step 2 of this section.

If no S-meter reading is obtained, the signal is not passing the RF and IF sections. Proceed to step 4.

4. Inject a modulated 455-kHz IF signal at the output of the second mixer (circuit 3). In a single-conversion receiver, the point of injection is the output of the first mixer, and the frequency may not be 455 kHz.

If audio output is obtained, the 455-kHz IF section is operating. Proceed to step 5.

If no audio output is obtained, the signal is not being passed through the 455-kHz IF section (circuit 4 or 5). Starting at the point of the signal injection and working toward the detector, measure the 455-kHz IF signal on the oscilloscope until a loss of signal is noted. This is the defective area.

5. Inject a modulated 7.8-MHz signal at the output of the first mixer (not all sets use 7.8-MHz first IF; use the appropriate frequency).

If audio output is obtained, the second oscillator (circuit 8) and second mixer (circuit 3) are probably good. Proceed to step 6.

If no audio output is obtained, either the second oscillator (circuit 8) or second mixer (circuit 3) is defective. Measure the output of the second oscillator on an RF voltmeter. If no RF is measured, the second oscillator is defective. If RF is present, the second mixer is defective. A defective AGC circuit (circuit 9) could bias the second mixer to cutoff. Measure AGC voltage with no RF signal applied.

6. Inject a modulated carrier frequency signal directly into the first mixer (bypass the RF amplifier).

If normal audio output is obtained, the trouble is in the RF amplifier (circuit 1) or AGC circuit (circuit 9).

If no audio is obtained, the first mixer (circuit 2) or AGC circuit (circuit 9) is suspect.

3-31. TROUBLESHOOTING PROCEDURE FOR "RADIO DOES NOT TRANSMIT OR RECEIVE" SYMPTOM

This procedure is used when there is no transmitter RF output and no receiver audio on any channel. The symptom can be verified by making transmitter RF power checks, and receiver audio power checks, on several channels. The symptom is valid if there is no measured power (receiver or transmitter) on any channel.

Other than the power supply, the only circuit that is common to the receiver and RF portion of the transmitter is the synthesizer (circuit 11). If the synthesizer fails, there is usually low receiver noise present. Of course, there is a remote possibility that there are two simultaneous failures, one in the receiver and one in the transmitter. The following steps will help isolate the trouble.

1. If there is any receiver noise, the trouble is probably in the synthesizer; go to step 2. If no noise is heard, go to step 5.

2. Check the output of the synthesizer with an RF voltmeter or oscilloscope, and a frequency counter. If no RF is measured, or if the RF is at the wrong frequency, the synthesizer is defective. Keep in mind that if one mixer input (say the synthesizer) is missing, there will still be RF output from the mixer, but at the wrong frequency. However, if the synthesizer output is absent or off-frequency, the mixer RF output level will be low. See step 4 for an alternative checkout method for the synthesizer.

3. A synthesizer typically contains at least two oscillators and a mixer. If any of these stages is defective, the correct output will not be generated. Measure RF output voltages at each stage until the defective stage is located, then make voltage and resistance checks to find the bad part in that stage.

4. An alternative to steps 2 and 3 is to use an RF signal generator, and inject signals of the correct frequency into the synthesizer (to substitute for the signals produced by the synthesizer oscillators). If the injected signal restores operation, the normal signal is missing from that point. Move the point of injection from the output of the synthesizer toward the input, one stage at a time. Each point may require a different frequency (refer to the service literature).

When the point of injection is moved beyond the defect, signal injection will no longer restore normal operation, and you have found the defective area. Keep in mind that this alternative procedure of signal injection requires a high-impedance probe to prevent circuit loading. A low-impedance probe used for injection can detune or disable the circuit and render the procedure useless.

5. If no receiver noise is heard during step 1, check the channel selector or meter illumination. If these lamps are on, input power is getting to the set. Check some of the major power distribution voltages (such as the 13.8-V distribution bus). If the lamps are not on, check input power and fuses.

6. If no problem is found in step 5, perform steps 2, 3, and 4.

7. If no other problem can be found, treat the "radio does not transmit" and "radio does not receive" symptoms as separate symptoms. Refer to the associated troubleshooting procedure for each of these symptoms (Secs. 3–30 and 3–35).

3–32. TROUBLESHOOTING PROCEDURE FOR "RADIO DOES NOT TRANSMIT OR RECEIVE ON SOME CHANNELS" SYMPTOM

This procedure is used to isolate a trouble when there is no transmitter RF power on certain channels and normal RF output on the other channels, and no received audio on certain channels and normal reception on the other channels.

This trouble is, in all probability, caused by failure of the synthesizer. Typically, the trouble results from a defective crystal, or possibly a faulty channel selector switch. These are the only parts involved in switching from one channel to another.

Troubleshooting the synthesizer is usually not difficult if you understand the synthesizer scheme (or how the synthesizer crystal-controlled oscillators are combined to produce the various channel frequencies). For example, Fig. 3–15 shows the classic synthesizer scheme for a typical 23-channel CB communications set. This scheme permits 10 crystals to generate all 23 channel frequencies. The crystals are connected in a 4 by 6 matrix to permit 24 possible frequency combinations. The channel selector switch is wired to select 23 of the 24 possible combinations. A crystal failure in such a synthesizer usually disables 4 or 6 channels.

A defective crystal is easy to identify from the synthesizer crystal frequency scheme. For example, if a set uses the scheme of Fig. 3–15, and channels 13, 14, 15, and 16 do not operate, crystal H is defective. If the problem is noted on channels 2, 6, 10, 14, 18, and 22, crystal B is the cause. Keep in mind that crystal frequencies are different with nearly every manufacturer, so you must know the synthesizer scheme (which should be included in the service literature).

Channel	Oscillator No. 1 crystal	Oscillator No. 2 crystal	Channel	Oscillator No. 1 crystal	Oscillator No. 2 crystal
1	A	E	13	A	H
2	B	E	14	B	H
3	C	E	15	C	H
4	D	E	16	D	H
5	A	F	17	A	I
6	B	F	18	B	I
7	C	F	19	C	I
8	D	F	20	D	I
9	A	G	21	A	J
10	B	G	22	B	J
11	C	G	–	C	J
12	D	G	23	D	J

Figure 3-15 Synthesizer scheme for a typical 23-channel CB communications set.

Also note that many present-day communications sets use PLL (phase-locked loop) instead of a synthesizer. This is particularly true with the 40-channel CB sets (and other multichannel communications sets). With PLL, all the frequencies are generated by two or three crystals and an integrated circuit (IC) which contains a programmable divider controlled by the channel selector. If any of the crystals or the IC fails in a PLL system, all the channels become inoperative (in most cases). Therefore, if you have a failure only on certain channels in a PLL system, the trouble is probably in the channel selector or associated wiring.

3-33. TROUBLESHOOTING PROCEDURE FOR "WEAK AUDIO" SYMPTOM

Use this troubleshooting procedure when the receiver audio is below rated power with a strong modulated carrier applied to the receiver. The symptom may also be accompanied by high distortion. Start by performing the PA mode check of Sec. 3-21.

If normal audio power output is possible in the PA mode, but not through the receiver with a modulated carrier applied, the trouble is in the first audio amplifier stage (circuit 12). Inject a modulated 455-kHz signal into the detector (circuit 6) and measure audio signal levels in circuit 12. Use an oscilloscope for the measurement and look for evidence of signal amplitude clipping or below-normal levels.

If audio power output is below rated power in the PA mode, as well as through the receiver, the trouble is in the audio amplifier circuits (circuits 15 through 18). Apply a 1-kHz signal to a loudspeaker, and use the loudspeaker signal to drive the microphone while operating in the PA mode. Make audio level checks using an oscilloscope in circuits 15 through 18. Look for evidence of signal amplitude clipping or below-normal levels.

If one of the push-pull audio power amplifiers (circuit 17 or 18) is disabled, audio power will drop to about 25% of normal, and the sound will be highly distorted.

3-34. TROUBLESHOOTING PROCEDURE FOR "POOR RECEIVER SENSITIVITY" SYMPTOM

Use this troubleshooting procedure when the receiver provides rated audio output, but does not meet the receiver sensitivity specification. This symptom is produced by a malfunction in the RF or IF amplifier section of the receiver (circuits 1 through 5).

1. Set up test equipment as for the receiver sensitivity check (Sec. 3-5), but increase the RF generator output until the 10-dB $(S + N)/N$ level is reached.

2. Check AGC voltage for normal values. If convenient, disable the AGC and recheck sensitivity. AGC can usually be disabled by shorting the AGC bus to chassis ground, or connecting a bias power supply to the AGC line, and setting the voltage for the normal no-signal AGC level. Consult the service literature regarding the AGC. If normal sensitivity is restored when AGC is disabled, troubleshoot the AGC circuit.

3. Measure the RF voltage levels of the synthesizer (circuit 11) and second oscillator (circuit 8) outputs. Low signal injection from the synthesizer can cause this symptom.

4. If the service literature includes typical stage gains, make gain measurements and compare them to the specified figures.

5. Inject 30% modulated 455 kHz at the detector (circuit 6) and adjust the RF generator output level for a convenient audio output level (such as 0.5 W).

6. Note the RF generator output level required to produce 0.5 W of audio output.

7. Move the point of injection toward the antenna, one stage at a time, and readjust the RF generator level for the same audio reference level (0.5 W). Change from 455 kHz to 7.8 MHz and then to the RF carrier frequency as the point of injection moves from the 455-kHz section toward the antenna.

8. Again note the RF generator output level.

9. The difference in the RF generator output levels between steps 6 and 8 is the gain (or loss) between injection points. If the RF generator output scale is

in microvolts, and the service literature stage gain is in decibels (or vice versa), convert the RF generator output readings as necessary.

10. Troubleshoot any stage with low gain, or any passive device (such as a coupling capacitor) with high attenuation, using voltage and resistance checks. If stage gain is low but there is no apparent defect in circuit components, try correcting the problem by realignment of the receiver, using the alignment procedures described in the service literature.

The procedure described in this section is a good troubleshooting technique even when the service literature does not give typical stage gains. Sometimes the defective stage is obvious (shorted capacitor, open resistor, etc.). Also, with experience, the typical stage gains will become memorized and a low stage gain will be easy to spot.

3-35. TROUBLESHOOTING PROCEDURE FOR "RADIO DOES NOT TRANSMIT" SYMPTOM

Use this troubleshooting procedure when there is no transmitter RF power output but receive operation is normal. The procedure also applies to the symptom in which the transmitter cannot be keyed. Start by checking if the transmit indicator lamp turns on when the push-to-talk switch is closed.

1. If the lamp does not turn on, place a jumper across the push-to-talk pins of the microphone jack or equivalent point in the set.

If the transmit indicator lamp now turns on (with the jumper), the problem is in the microphone or microphone cord. Try replacing the microphone and cord.

If the transmit indicator still does not turn on, the transmit-receive relay is probably the cause. In many present-day sets, the transmit-receive relay is replaced by a solid-state switching circuit. Typically, these switching circuits are made up of diodes that alternately connect the transmit and receive circuits when alternately forward-biased and reverse-biased. Solid-state switching circuits are generally more difficult to troubleshoot than a transmit-receive relay.

2. If the transmit indicator does turn on, but there is no RF transmission, the problem can be in almost any of the transmitter RF circuits (circuits 21 through 25). Use an RF voltmeter or high-frequency oscilloscope to measure RF voltages. Starting with circuit 21 and working toward circuit 25, measure each accessible point in the signal path until the absence of RF voltage is noted. This is the defective area.

In some sets, circuits 21 and 22 may be part of the synthesizer (or a PLL), and the synthesizer output in the transmit mode is a low-level RF carrier frequency. If the set uses such a design, start by checking the RF from the synthesizer and then go on to circuits 23 through 25.

3-36. TROUBLESHOOTING PROCEDURE FOR "LOW TRANSMITTER POWER" SYMPTOM

Use this troubleshooting procedure if the transmitter RF output power is below normal, and receiver operation is normal. The symptom is caused by low gain or low voltage in the transmitter RF amplifiers (circuits 23 through 25). Measure RF voltages at each of these amplifiers. Follow this with voltage and resistance checks. If there is no apparent fault, but RF power is low, try correcting the problem by alignment of the RF amplifiers, using the procedures described in the service literature. Usually, the RF amplifiers only require "peaking" of the tuned circuits to get proper alignment.

3-37. TROUBLESHOOTING PROCEDURE FOR "RADIO OFF-FREQUENCY" SYMPTOM

Use this troubleshooting procedure if the transmitter frequency is not within specification on any channel or channels. The receiver may also operate off-frequency, but detection of the symptom is more difficult. An off-frequency receiver displays symptoms of poor sensitivity (and possibly distortion). Normal results may be obtained during test because the RF generator is usually tuned for maximum receiver output, not necessarily to the channel frequency.

If the set is off-frequency, the problem is usually one of the crystal oscillators. If the trouble appears on all channels, the transmit oscillator (circuit 21) is operating off-frequency. If the trouble appears only on certain channels, refer to the procedures of Sec. 3-32. However, instead of certain channels being inoperative, there will be certain channels that are off-frequency. Replace the crystals or tuning components (coils/capacitors) that may detune the crystal operating frequency. Slight off-frequency conditions may be improved by realignment.

3-38. TROUBLESHOOTING PROCEDURE FOR "ABNORMAL TRANSMITTER MODULATION" SYMPTOM

Use this troubleshooting procedure if there is no transmitter modulation, and receiver audio power is normal. Also use the procedure if the transmitter appears undermodulated, or if the transmitter is easily overmodulated.

If there is no modulation, the microphone, microphone amplifier (circuit 14), or modulation input to the transmitter RF power amplifier (circuit 25) is defective.

1. Perform the PA mode check (Sec. 3-21).

2. If PA mode operation is normal, troubleshoot the audio output transformer and audio input to circuit 25.

3. If PA mode does not produce audio output, inject a 1-kHz test signal into the audio pins of the microphone jack (or equivalent points in the set) and key the transmitter.

4. If you get modulation, the microphone or microphone cord is faulty.

5. If no modulation is produced, troubleshoot the microphone amplifier (circuit 14).

6. If an unusually high audio level is required for 100% modulation (or you can not get 100% modulation), perform the PA mode check. If rated audio power is possible in the PA mode, troubleshoot the audio input portion of circuit 25. If low audio power is measured on the PA mode check, troubleshoot the microphone amplifier (circuit 14) for low gain, and the microphone for low output (try a substitute microphone, if available).

7. If overmodulation is present, troubleshoot the modulation limiting circuitry, which is usually part of circuit 25.

3–39. TROUBLESHOOTING PROCEDURE FOR "ABNORMAL SSB TRANSMITTER OPERATION" SYMPTOM

Use this troubleshooting procedure for any abnormal SSB transmit condition (no output, low power, improper modulation, or incorrect frequency), including symptoms in which the SSB transmitter and receiver both show abnormal operation.

1. If there is no SSB transmitter RF power output, check the SSB receiver audio power.

2. If both modes are inoperative, the trouble could be in the ring (balanced) modulator/product detector (circuit 28) or narrowband 7.8-MHz bandpass filter (circuit 30) shown in Fig. 3–11. Apply a two-tone test signal to the microphone and measure the RF output of the ring modulator (circuit 28). If no output is measured, troubleshoot the ring modulator circuit. If output is measured, troubleshoot the 7.8-MHz bandpass filter.

3. If receiver operation is normal, the SSB transmitter RF amplifiers are the suspected stages (circuits 31 and 32). Apply a two-tone test signal to the microphone and measure RF voltages in circuits 31 and 32. Troubleshoot the area where RF voltage is first missing.

4. If there is no transmitter and receiver operation on one sideband only, the problem is likely in the 7.8015/7.7985-MHz oscillator (circuit 21). This circuit uses one crystal in the AM and one sideband mode, and another crystal in the opposite sideband mode. Troubleshoot the crystal and mode selector switch.

5. If transmitter RF power is low, troubleshoot the transmitter RF amplifiers for low gain (circuits 31 and 32) as described in Sec. 3–36.

3-40. TROUBLESHOOTING PROCEDURE FOR "ABNORMAL SSB RECEIVER OPERATION" SYMPTOM

Use this troubleshooting procedure for any abnormal SSB receive condition (no output, poor sensitivity, or poor adjacent sideband rejection). The AM mode and SSB transmit modes are normal.

1. If there is no receiver audio or poor sensitivity, set up test equipment for an SSB receiver sensitivity check (Sec. 3–16).

2. Disable the SSB AGC circuit (circuit 27). If normal operation is restored, troubleshoot the AGC circuit.

3. Troubleshoot the 7.8-MHz IF amplifier (circuit 29).

4. If adjacent sideband rejection does not meet specification, troubleshoot the narrow-band 7.8-MHz bandpass filter (circuit 30).

3-41. TROUBLESHOOTING PROCEDURE FOR "RECEIVER DISTORTION" SYMPTOM

Use this procedure when receiver audio does not meet the distortion specification or there is a symptom of distorted audio in the AM mode. Start by measuring distortion in the PA mode.

1. If distortion is measured in the same degree in the PA mode as in the receive mode, measure audio signals in circuits 15 through 18 using an oscilloscope. Starting at circuit 15 and working toward the loudspeaker, observe waveforms for a point where a change in the waveform occurs.

2. If distortion meets specification in the PA mode, check distortion with a 1-kHz test signal injected at the detector.

If the same degree of distortion is measured, measure audio waveforms in circuit 12, and look for the point where the waveform changes.

If distortion meets specification, check the receiver for off-frequency condition, lack of AGC action, and IF amplifier distortion.

3-42. TROUBLESHOOTING PROCEDURE FOR "ABNORMAL PA MODE OPERATION" SYMPTOM

Use this troubleshooting procedure only when the PA mode is inoperative but all other modes are normal. With a set as shown in Figs. 3–10 and 3–11, this symptom can be caused only by faulty PA jack wiring or the PA mode select switch.

3-43. TROUBLESHOOTING PROCEDURE FOR "ADJACENT CHANNEL INTERFERENCE" SYMPTOM

Use this troubleshooting procedure when the receiver does not meet the adjacent channel rejection specification in the AM mode, but all other performance specifications are normal.

1. Recheck receiver alignment as described in the service literature.

2. Check the bandpass filter, and all tuned circuits in the RF and IF stages as follows:

Inject a modulated 455-kHz signal at the output of the second mixer. Tune the RF generator to 10 kHz above and below the 455-kHz center frequency. Measure the output of the 455-kHz bandpass filter (circuit 4) as the RF generator is tuned across the band. The bandpass should be symmetrical. A defective part usually causes a nonsymmetrical condition (but severe misalignment can also result in nonsymmetrical response in tuned circuits).

Measure the bandpass at each accessible point in the signal path through the 455-kHz IF amplifier (circuit 5).

If the 455-kHz circuits show symmetrical bandpass, check the bandpass of the 7.8-MHz IF circuits (circuits 1 and 3).

If the receiver meets specifications, little can be done to further improve adjacent channel rejection.

3-44. ANTENNA AND TRANSMISSION-LINE MEASUREMENTS WITH LIMITED TEST EQUIPMENT

In general, antennas and transmission lines (lead-ins) used with radio communications sets are best tested during troubleshooting by means of commercial SWR meters, field strength meters, and the special test sets described in Chapter 2. However, it is possible to make a number of significant tests using basic meters (voltmeter, ohmmeter, ammeter). The following paragraphs describe how these procedures can be performed when commercial antenna test devices are not available to the troubleshooter.

3-44.1 Antenna Length and Resonance Measurements

Most antennas are cut to length related to the wavelength of the signal being transmitted or received. Generally, antennas are cut to one-half wavelength (or one-quarter wavelength) of the center operating frequency. The electrical length of an antenna is always greater than the physical length, due to capacitance and end effects. Therefore, two sets of calculations are required: one for electrical length and one for physical length. The calculations for antenna length and resonant frequency are shown in Fig. 3-16.

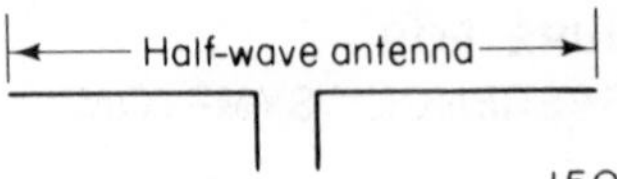

|← —— Half-wave antenna —— →| Basic Hertz type

$$\text{Meters} = \frac{150}{\text{Frequency (MHz)}}$$

Electrical
length

$$\text{Feet} = \frac{492}{\text{Frequency (MHz)}}$$

$$\text{Inches} = \frac{5906}{\text{Frequency (MHz)}}$$

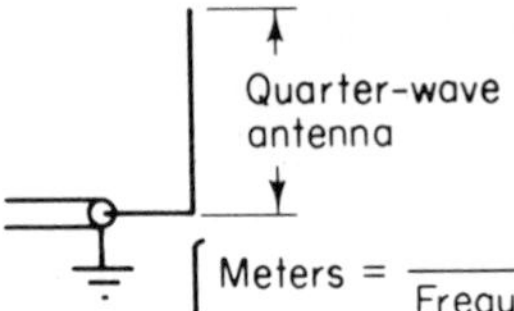

Quarter-wave
antenna   Basic Marconi type

$$\text{Meters} = \frac{75}{\text{Frequency (MHz)}}$$

Electrical
length

$$\text{Feet} = \frac{246}{\text{Frequency (MHz)}}$$

$$\text{Inches} = \frac{2953}{\text{Frequency (MHz)}}$$

Physical length (approx.) = electrical length x K factor

K factor = 0.96 for frequencies below 3 MHz
= 0.95 for frequencies between 3 and 30 MHz
= 0.94 for frequencies above 30 MHz

Figure 3-16   Calculations for antenna length.

3-44.2 Practical Resonance Measurements for Antennas

With a short antenna it is possible to measure the exact physical length and find the electrical length (and the resonant frequency) using the equations of Fig. 3-16. Obviously, this is not practical for long antennas. Also, the exact resonant frequency (electrical length) may still be in doubt for short antennas due to the uncertain K factor of Fig. 3-16. Therefore, for practical purposes the electrical length and resonant frequency of an antenna should be determined electrically.

There are three practical methods for determining antenna resonant frequency: dip adapter circuit, antenna ammeter, and wavemeter.

Basic dip adapter antenna resonance measurement. The dip adapter circuit described in Sec. 2-10 can be used to measure antenna resonance. The basic technique is to couple the adapter to the antenna as if the antenna is a resonant circuit, tune for a tip, and read the resonant frequency. However, there are certain precautions to be observed.

The measurement can be made as a conventional resonant circuit provided that the antenna is accessible, allowing the dip adapter to be coupled directly to the antenna elements. If the antenna is a simple grounded element (no matching problems between antenna and transmission line), the adapter can be coupled to the transmission line. However, if the antenna is fed by a coaxial line or any system in which the line is matched to the antenna (which is usually the case for communications equipment), it is best to couple the adapter coil directly to the antenna elements (to get the most accurate results).

No matter how carefully the antenna and lead-in are matched, there is some mismatch, at least over a range of frequencies. This means that there are two reactances (or impedances) that interact to produce extra resonances. If resonance is measured under such conditions, a dip is produced at the correct antenna frequency (or at harmonics), and another dip (plus harmonics) at the extra frequency. It can sometimes be very confusing to tell the dips apart. Also, antenna resonance measurements should be made with the antenna in the actual operating position. The nearness or directive or reflective elements, as well as the height, affects antenna characteristics and possibly changes resonant frequency.

The dip-adapter procedure for grounded antennas is as follows:

1. Couple the dip adapter to the antenna tuner if the antenna is tuned, as shown in Fig. 3-17.

2. If the antenna is untuned or it is not practical to couple to the antenna tuner, disconnect the antenna lead-in and couple the lead-in to the adapter through a pickup coil, as shown in Fig. 3-18.

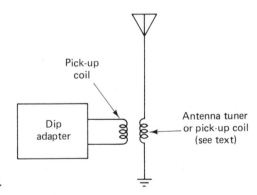

Figure 3-17 Basic dip-adapter circuit.

To antenna

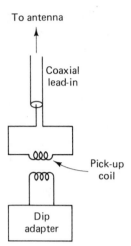

Figure 3-18 Dip-adapter measurement of antenna resonant frequency.

3. Set the generator to its lowest frequency. Adjust the signal generator output for a convenient reading on the adapter meter.

4. Slowly increase the generator frequency, observing the meter for a dip indication. Tune for the bottom of the dip.

5. Note the frequency at which the first (lowest frequency) dip occurs. This should be the primary resonant frequency of the antenna. As the signal generator frequency is increased, additional dips should be noted. These are harmonics and should be exact multiples of the primary resonant frequency. Check two or three of these frequencies to make sure that they are harmonics. Then go back to the lowest frequency dip to ensure that the lowest frequency is the primary resonant frequency.

The dip-adapter procedure for ungrounded antennas is as follows:

1. Disconnect the antenna lead-in or feed line from the antenna.

2. If the antenna is center-fed, short across the feed point with a piece of wire.

3. Couple the dip-adapter coil directly to the antenna. Usually, the best results are obtained by coupling at a maximum current (low-impedance) point. For example, the maximum current point occurs at the center of a half-wave antenna.

4. Starting at the lowest signal generator frequency and working upward, tune the signal generator for a dip on the meter as described for grounded antennas. The lowest dip is the primary resonant frequency.

Series ammeter measurement of the antenna resonant frequency.   A series ammeter can be used to find the resonant frequency of an antenna. The basic circuit is shown in Fig. 3–19. The signal generator is tuned for a maximum reading on the ammeter, indicating a maximum transfer of energy from the

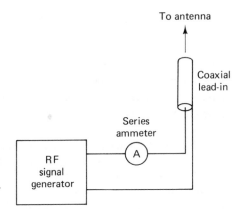

Figure 3-19 Series ammeter measurement of antenna resonant frequency.

generator into the antenna (as a result of both being at the same frequency). The antenna frequency is then read from the generator dial or frequency counter (if used). The series ammeter method has the advantage of measuring the combined resonant frequency of both the antenna and the transmission line. This is most practical since in normal communications operation the antenna is operated with the transmission line.

A version of the series ammeter method is often used in transmitters as an indicator for antenna tuning. Most transmitters are crystal-controlled and operated at a specific frequency with the antenna tuned to that frequency. The electrical length (and consequently the resonant frequency) of the antenna is varied by a reactance in series with the lead-in, as shown in Fig. 3-20. The reactance can be a variable capacitor or variable inductance. With such an arrangement, the transmitter is tuned to its operating frequency, and then the antenna

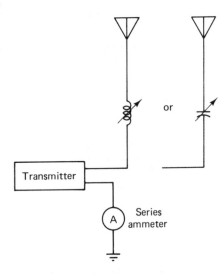

Figure 3-20 Tuning an antenna to resonant frequency of associated transmitter using the series ammeter method.

is tuned to that frequency as indicated by a maximum reading on the series ammeter.

The series ammeter method has certain drawbacks, one being the operating frequency limit of the series ammeter. Another is the fact that the series ammeter consumes some power in operation. However, the series ammeter has an advantage in that true antenna power can be calculated (as discussed in Sec. 3–44.3).

Wavemeter measurement of the antenna resonant frequency.   A wavemeter can be used to find the resonant frequency of an antenna. The basic wavemeter circuit, shown in Fig. 3–21, is essentially a tuned resonant circuit, detector, and indicator. A commercial wavemeter has a precision-calibrated tuning dial so that exact frequency can be measured, and an amplifier circuit so that weak signals can be measured. When used to measure antenna resonant frequency, the wavemeter is tuned to the approximate resonant frequency of the antenna, and the signal generator is tuned for a maximum reading on the wavemeter, and then the generator frequency is measured on the frequency counter. In this case, the wavemeter resonant circuit is broadly tuned. When used to tune an antenna, the wavemeter is tuned to the approximate transmitter operating frequency, and then the antenna is tuned for a maximum reading on the wavemeter. Not all wavemeters are provided with precision tuning. Such

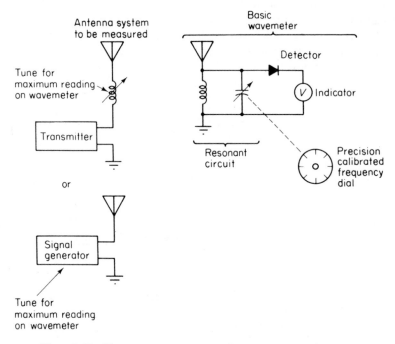

Figure 3–21   Wavemeter measuremeter of antenna resonant frequency.

wavemeters serve only as a maximum (or peak) readout device, similar to the field strength meter (Sec. 2-8).

3-44.3 Antenna Impedance and Radiated Power Measurements

The impedance of an antenna is not constant along the entire length of the antenna. In a typical half-wave antenna as shown in Fig. 3-22, the impedance is minimum at the center and maximum at the ends. In theory, the impedance is zero at the center. Since the antenna is fed at some point away from the exact center, there is some impedance for any antenna. A typical antenna used in mobile communications sets has an impedance of 50 to 72 Ω, whereas a half-wave TV antenna has 300 Ω.

Antenna impedance is determined using the basic Ohm's law equation $Z = E/I$, with voltage and current being measured at the antenna feed point. However, such measurements are not usually made in practical applications.

Radiation resistance is a more meaningful term. When the d-c resistance of the antenna is disregarded (antenna d-c resistance is usually a few ohms or less, except in very low-frequency, long-wire antennas), the antenna impedance can be considered as the radiation resistance. Radiated power can then be determined using the basic Ohm's law equation $P = I^2R$.

Practical impedance and radiated power measurement for antennas. On those antennas designed to be used with coaxial transmission lines (as is the case with most communications sets), the antenna and transmission-line impedance must be matched. In this case, the impedance match between transmis-

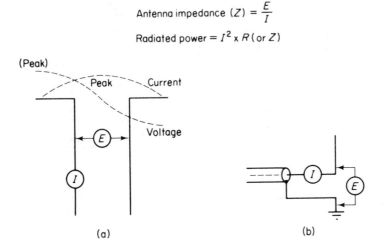

Figure 3-22   Theoretical calculations for antenna impedance and radiated power: (a) half-wave; (b) quarter-wave.

sion line and antenna is of greater importance than actual impedance value (both antenna and transmission line must be 50 Ω, 72 Ω, etc.). The condition of match (or mismatch) between antenna and transmission line can best be tested by the SWR measurement discussed in Chapter 2 and in this chapter.

The following procedure can be used to find the impedance and radiated power of any antenna system. However, it should be noted that the impedance and power obtained are for the complete antenna system (antenna and transmission line), as seen from the measurement end (which is generally where the transmission line connects to the communications set antenna connector).

1. Connect the equipment as shown in Fig. 3-23. Disconnect the transmission line from the antenna terminal of the communications set, and then reconnect the transmission line to the signal generator with the series ammeter in place as shown. The transmitter of the communications set can be used in place of the signal generator. However, such transmitters usually operate on fixed frequencies, thus limiting the measurements to those frequencies.

2. Adjust the signal generator to the center frequency at which the antenna is used (or for any desired operating frequency to which the antenna can

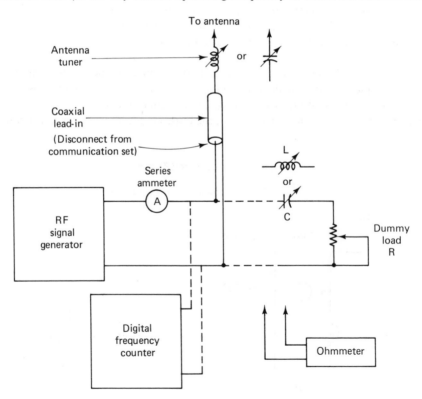

Figure 3-23 Measuring antenna impedance and radiated power using resistance substitution (ammeter method).

be tuned). If the transmitter is used in place of the signal generator, operate the transmitter at (or near) the center frequency.

3. If the antenna can be tuned, tune the antenna for a maximum indication on the ammeter. Record the indicated current. If the antenna cannot be tuned, adjust the signal generator maximum on the ammeter, and record the current. If the antenna cannot be tuned, and you are using a fixed-frequency transmitter, simply record the indicated current.

4. If a digital frequency counter or precision wavemeter is available, verify that the signal generator (or transmitter) and antenna are tuned to the correct frequency (after the antenna and/or generator have been tuned for maximum indication on the ammeter).

5. Disconnect the transmission line from the series ammeter and signal generator. Connect the signal generator to the dummy load through the series ammeter as shown. If the antenna was tuned in step 3, adjust inductance L or capacitor C for maximum indication on the ammeter. (If the antenna is tuned by means of a variable inductance, use capacitor C to tune the dummy load, and vice versa). If the antenna is not tuned, adjust the signal generator for maximum indication. (If the dummy load is actually noninductive, and is a pure resistance, it should not be necessary to readjust the signal generator from the frequency in step 3 when the dummy load is connected.)

6. Adjust the dummy-load resistance until the indicated current on the series ammeter is the same as the antenna current recorded in step 3.

7. Remove power from the circuit. Measure the d-c resistance of R with an ohmmeter. This resistance is equal to the antenna system impedance (or radiation resistance), at the frequency of measurement.

8. Calculate the actual power delivered to the antenna (or radiated power) using:

$$\text{radiated power} = I^2 \times R \,(\text{or } Z)$$

where I = indicated current (amperes)
R = radiation resistance (or antenna system impedance)

An alternative method must be used when the operating frequency is beyond the range of the available ammeter, when no ammeter is available, or when the ammeter presents an excessive load. A precision 1-Ω, noninductive resistor and voltmeter can be used in place of the ammeter, as shown in Fig. 3-24. With a 1-Ω resistor, the indicated voltage is equal to the current passing through the resistor (and antenna system). Except for the connections, the procedure is identical to the one that requires an ammeter.

3-45. OSCILLATOR TROUBLESHOOTING

Most communications equipment uses some form of oscillator for signal generation and frequency control. For example, the frequency-control circuits of a CB set can be quite simple or fairly complex, depending on the set. In the

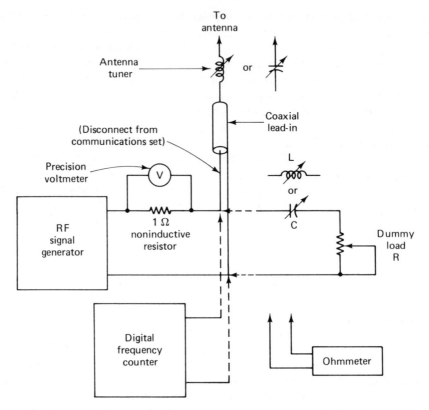

Figure 3-24 Measuring antenna impedance and radiated power using resistance substitution (voltmeter method).

simplest form, such as a hand-held CB, there are two oscillators, one for transmitter frequency control and one for the receiver local oscillator. In a "typical" communications set there are two or three oscillators, combined with one or more mixers, to form a *frequency synthesizer* that produces transmitter and receiver local oscillator signals. A PLL (phase-locked-loop) set contains a standard oscillator (probably within the PLL IC, but having an external crystal) and one or two other oscillators to form the complete frequency synthesizer.

No matter how complex the circuits appear, they are essentially oscillators, and can be treated as such from a practical troubleshooting standpoint. That is, each circuit contains oscillators, which produce signals (probably crystal-controlled). These signals must have a given amplitude and must be at a given frequency (or capable of tuning across a given frequency range) for the set to operate properly. Thus, if you measure the signals and find them to be of the correct frequency and amplitude, the oscillators are good from a troubleshooting standpoint.

3-45.1 Oscillator Test and Troubleshooting Procedures

The first step in troubleshooting any oscillator circuit is to measure both the amplitude and frequency of the output signal. Many oscillators have a built-in test point. If not, the signal may be monitored at the collector or emitter (plate or cathode for a vacuum-tube oscillator), as shown in Fig. 3-25. Signal amplitude is monitored with a meter or oscilloscope using an RF probe. The simplest way to measure oscillator signal frequency is with a frequency counter.

Oscillator frequency problems.   When you measure the oscillator signal, the frequency is (1) right on, (2) slightly off, or (3) way off.

If the frequency is slightly off, it is possible to correct the problem with adjustment. Most oscillators are adjustable, even those that are crystal-controlled. Usually, the RF coil or transformer is slug-tuned. The most precise adjustment is obtained by monitoring the oscillator signal with a frequency counter and adjusting the circuit for exact frequency. However, it is also possible to adjust an oscillator using a meter or oscilloscope.

When the circuit is adjusted for maximum signal amplitude the oscillator is at the crystal frequency. However, it is possible (but not likely) that the oscillator is being tuned to a harmonic (multiple or submultiple) of the crystal

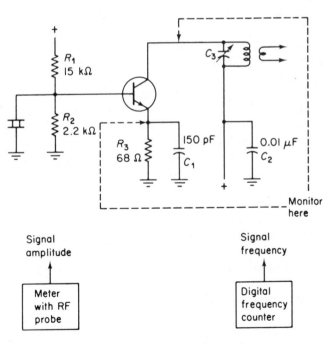

Figure 3-25 Oscillator signal test points.

frequency. The frequency counter shows this, whereas a meter or oscilloscope does not.

If the oscillator frequency is way off, look for a defect rather than improper adjustment. For example, the coil or transformer may have shorted turns, the transistor or capacitor may be leaking badly, or the wrong crystal is installed in the right socket (this does happen).

Oscillator signal amplitude problems.   When you measure the oscillator signal, the amplitude is (1) right on, (2) slightly low, or (3) very low.

If the amplitude is slightly low, it is possible to correct the problem with adjustment. Monitor the signal with a meter or oscilloscope, and adjust the oscillator for maximum signal amplitude. This also locks the oscillator on the correct frequency. If adjustment does not correct the problem, look for leakage in the transistor, or for a transistor with low gain.

If the amplitude is very low, look for defects such as low power-supply voltages, badly leaking transistors and/or capacitors, and shorted coil or transformer turns. Usually, when signal output is very low, there are other indications, such as abnormal voltage and resistance values.

Oscillator bias problems.   One of the problems in troubleshooting solid-state oscillator circuits is the bias arrangement. RF oscillators are generally reverse-biased so that they conduct on half-cycles. However, the transistor is initially forward-biased by d-c voltages (through the bias networks, as shown in Fig. 3–26. This turns the transistor on so that the collector circuit starts to conduct. Feedback occurs, and the transistor is driven into heavy conduction.

During the time of heavy conduction, a capacitor connected to the transistor base is charged in the forward-bias direction. When saturation is reached, there is no further feedback, and the capacitor discharges. This reverse-biases the transistors, and maintains the reverse bias until the capacitor has discharged to a point where the fixed forward bias again causes conduction.

This condition presents a problem in the operation of class C solid-state RF oscillators. If the capacitor is too large, it may not discharge in time for the next half-cycle. In that case, the class C oscillator acts as a blocking oscillator, controlling the frequency by the capacitance and resistance of the circuit (instead of by the capacitance and inductance, as should be the case for RF oscillators). If the capacitor is too small, the class C oscillator may not start at all. This same condition is true if the capacitor is leaking badly. From a practical troubleshooting standpoint, the measured condition of bias on a solid-state oscillator can provide a good clue to operation if you know how the oscillator is supposed to operate.

The oscillator in Fig. 3–26 is initially forward-biased through R1 and R3. As Q1 starts to conduct and in-phase feedback is applied to the emitter (to sustain oscillation), capacitor C1 starts to charge. When saturation is reached (or

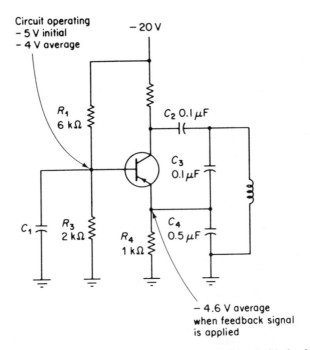

Figure 3-26   Class C RF oscillator (reverse-biased or zero-biased with circuit operating).

approached) and the feedback stops; capacitor C1 then discharges in the opposite polarity, reverse-biasing Q1. The value of C1 is selected so that C1 discharges to a voltage less than the fixed forward bias before the next half-cycle. Thus, transistor Q1 conducts on slightly less than the full half-cycle. Typically, a class C RF oscillator such as the one shown in Fig. 3-26 conducts on about 140° of the 180° half-cycle.

Exploring the subject of bias further, it is commonly assumed that transistor junctions (and diodes) start to conduct as soon as forward voltage is applied. This is not true. Figure 3-27 shows characteristic curves for three different types of transistor junctions. All three junctions are silicon, but the same condition exists for germanium junctions. None of the junctions conduct noticeably at 0.6 V, but current starts to rise at that point. At 0.8 V, one junction draws almost 80 mA. At 1 V, the d-c resistance is on the order of 2 or 3 Ω, and the transistor draws almost 1 A. In a germanium transistor, noticeable current flow occurs at about 0.3 V.

For troubleshooting purposes, bias measurements provide a clue to the performance of solid-state oscillators, although such measurements do not provide positive proof. The one sure test of an oscillator is to measure output signal amplitude and frequency.

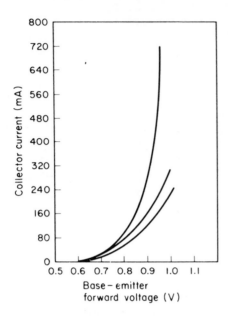

Base – emitter
forward voltage (V)

Figure 3–27 Characteristic curves for silicon transistor junctions.

Oscillator quick-test. It is possible to check whether an oscillator circuit is oscillating by using a voltmeter and a large-value capacitor (typically 0.01 μF or larger). Measure either the collector or emitter voltage with the oscillator operating normally, and then connect the capacitor from base to ground as shown in Fig. 3–28. This should stop oscillation, and the emitter or collector voltage should change. When the capacitor is removed, the voltage should

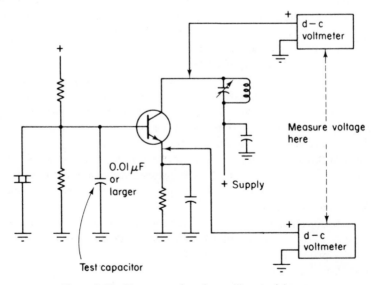

Figure 3–28 Test connections for oscillator quick-test.

return to normal. If there is no change when the capacitor is connected, the oscillator is probably not oscillating. In some oscillators, you get better results by connecting the capacitor from the collector to ground. Also, do not expect the voltage to change on an element without a load. For example, if the collector is connected directly to the supply voltage, or through a few turns of wire as shown in Fig. 3-28, this voltage does not change, with or without oscillation.

4

Digital

Test

Equipment

This chapter describes both general and special test equipment used for digital troubleshooting.

4-1. OSCILLOSCOPES FOR DIGITAL TROUBLESHOOTING

An oscilloscope for digital troubleshooting should have a bandwidth of at least 50 MHz and preferably 100 MHz. The pulse width or pulse durations found in most digital equipment are on the order of a few microseconds, often only a few nanoseconds wide.

A *triggered horizontal sweep* is an essential feature. Preferably, the delay introduced by the trigger should be very short. Often, the pulse to be monitored occurs shortly after an available trigger. In other cases, the horizontal sweep must be triggered by the pulse to be monitored. In some oscilloscopes, a delay is introduced between the input and the vertical deflection circuits. This permits the horizontal sweep to be triggered before the vertical signal is applied, thus assuring that the complete pulse is displayed.

Dual-trace or *multiple-trace* sweeps are also essential. Most trouble-shooting is based on the monitoring of two time-related pulses (say an input pulse and output pulse, or a clock and a readout pulse). With dual trace, the two pulses can be observed simultaneously, one above the other or superimposed if convenient. In other test configurations, a clock pulse is used to trigger both horizontal sweeps. This allows for a three-way time relationship measurement (clock pulse and two circuit pulses, such as one input and one output). A few oscilloscopes have multiple-trace capabilities. However, this is not standard. Usually, such an oscilloscope is provided with plug-in options that increase the number of horizontal sweeps.

The sensitivity of both the vertical and horizontal channels should be such that full-scale deflection can be obtained (without overdriving or distortion) with less than a 1-V signal applied. Typically, the signal pulses used in digital work are on the order of 5 V, but often they are less than 1 V (such as when ECL is involved).

In some cases, *storage oscilloscopes* (for display of transient pulses) and *sampling oscilloscopes* (for display and measurement of very short pulses) may be required. However, most digital work can be done with an oscilloscope having the bandwidth and sweep capabilities just described.

4-2. MULTIMETERS FOR DIGITAL TROUBLESHOOTING

The voltmeters for digital troubleshooting should have essentially the same characteristics as those for other solid-state troubleshooting. As usual, meters with digital readouts are easier to use than meters with analog (moving-needle) readouts. Generally, a very high input impedance is not critical for digital work (200,000 Ω/V is usually sufficient).

Typically, digital circuit operating voltages are 12 V, while the logic voltages (pulse levels) are 5 V or less. Thus, the voltmeter should have good resolution on the low-voltage scales. Here again, the digital readout meter has the advantage. For example, a typical logic level might be where 0 is represented by 0 V, and 1 is represented by 3 V. This means that an input of 3 V or greater to an OR gate will produce a 1 (or true) condition, while an input of something less than 3, say 2 V, will produce a 0 (false) condition. As is typical for digital logic specifications, the region between 2 and 3 V is not spelled out. If the voltmeter is not capable of an accurate readout between 2 and 3 V on some scale, you can easily arrive at a false conclusion. If the OR gate in question is tested by applying a supposed 3 V, which is actually 2.7 V, the OR gate may or may not operate to indicate the desired 1 (or true) output.

Generally, d-c voltage accuracy should be ±2% or better, with ±3 to ±5% accuracy for the a-c scales. The a-c scales will probably be used only in checking power supply or input power functions in digital equipment, since all signals are in the form of pulses (requiring an oscilloscope for display and measurement).

The ohmmeter portion of the multimeter should have the usual high-resistance ranges. Many of the troubles in the most sophisticated and complex computers boil down to such common problems as cold solder joints, breaks in printed-circuit wiring that result in high resistance, or shorts or partial shorts between wiring (producing an undesired high-resistance condition). The internal battery voltage of the ohmmeter should not exceed any of the voltages used in the circuit being checked.

4-3. PULSE GENERATORS FOR DIGITAL TROUBLESHOOTING

Ideally, a pulse generator should be capable of duplicating any pulse present in the circuits being tested. Thus, the pulse generator output should be continuously variable (or at least adjustable by steps) in amplitude, pulse duration (or width), and frequency (or repetition rate) over the same range as the circuit pulses. That is not always possible in every case. However, with modern laboratory pulse generators, it is generally practical for most digital equipment. Typically, the pulses are ± 5 V or less in amplitude, but could be 10 to 15 V in rare cases (such as when CMOS is operated with a high-voltage supply). The pulses are rarely longer than 1 s or shorter than 1 ns, although there are exceptions here, too. Repetition rates are generally less than 100 kHz, but could run up to 10 MHz (or higher).

Some pulse generators have special features, such as two output pulses with variable delay between the pulses or an output pulse that can be triggered from an external source. However, most routine digital test and troubleshooting can be performed with standard pulse generators.

4-4. LOGIC PROBE

Figure 4–1 shows a typical logic probe. Such logic probes are used to monitor in-circuit logic activity. By means of a simple lamp indicator, a logic probe tells you the logic state of a digital signal and allows brief pulses to be detected. Logic probes detect and indicate high and low (1 or 0) logic levels, as well as intermediate or "bad" logic levels, including an open circuit, on a single line of a digital circuit. For example, the probe can be connected to one of the address or data bus lines (or the clock line, chip-select line, etc.) and will indicate the state (0, 1, or high-impedance) of that line. Note the extender clamp connected to an IC at the left of the probe in Fig. 4–1. Such extenders are used where the IC terminals are not readily accessible for connection of test leads.

A logic-level indicator lamp, near the probe tip, gives an immediate indication of the logic states, either static or dynamic, existing in the circuit under test when the probe is touched to the circuit or line. Figure 4–2 shows how the logic-level indicator lamp responds to voltage levels and pulses found on a typical TTL digital circuit line. The indicator lamp can give any of four indications: (1) off, (2) dim (about half brilliance), (3) bright (full brilliance), or (4) flashing on and off. The lamp is normally in its dim state, and must be driven to one of the other three states by voltage levels at the probe tip. The

Figure 4-1 Hewlett-Packard 545A logic probe. (Courtesy Hewlett-Packard Company.)

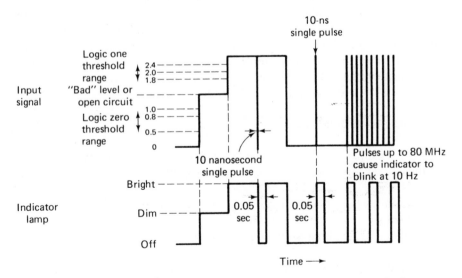

Figure 4-2 How the logic-level indicator lamp responds to voltage levels and pulses found on a typical TTL digital circuit line.

lamp is bright for inputs at or above the 1 state and off for inputs at or below 0. The lamp is dim for voltages between the 1 and 0 states and for open circuits. Pulsating inputs cause the lamp to flash at about a 10-Hz rate.

The logic probe is particularly effective when used with the logic pulses described in Sec. 4–5. For example, the logic pulser can be used to simulate the input to an element in the digital circuit (such as a clock input to a microprocessor, RAM, ROM, or I/O), while the logic probe is touching the output of the element to sense activity of the element. In summary, a logic probe is a self-contained, easy-to-use tool for examining logic nodes. You can use the probe to verify logic continuity, signal flow, address decoding, clock, switch, and bus device activity. You can also use a logic probe to check digital circuit operating characteristics in the *single-step mode*, as described in Sec. 4–9.

4–4.1 Using the Logic Probe in Digital Circuits

The following paragraphs provide brief descriptions of how the logic probe can be used in troubleshooting. More detailed procedures and approaches are given in the remaining sections of this chapter, and in Chapter 5.

4–4.2 Probe Power Supply

The logic probe can be powered from the digital equipment power supply or from a regulated d-c power supply. If a separate power supply is used, the power supply and digital equipment grounds should be connected together. The power-supply voltage range for TTL operation is 4.5 to 15 V. A ground wire (provided with the probe) may be connected just behind the probe indicator window. The ground wire is a convenient means of connecting grounds when using external regulated power supplies. The ground wire also improves pulse-width sensitivity and noise immunity. However, use of the ground wire is optional and is not required for all applications.

4–4.3 Pulse Detection

The logic probe is ideal for detecting short-duration and low-repetition-rate pulses that are difficult to observe on an oscilloscope. Typically, positive pulses of about 10 ns or greater trigger the indicator on for at least 50 ms. Negative pulses cause the indicator to go off momentarily.

4–4.4 Testing Three-State Logic Outputs

The bad-level feature of the logic probe is useful for testing three-state logic outputs. The logic-high and logic-low states are detected as described under pulse detection (Sec. 4–4.3). The third state found in many microcomputer circuits (the high-impedance output) is detected as an open-circuit (or bad-level) condition, which leaves the probe indicator dim. The bad-level indicator is also useful for detecting floating or unconnected TTL inputs, which look like a bad level to the logic probe.

4-4.5 Basic Probe Techniques

Several logic circuit analysis techniques are useful with the logic probe, as discussed in this chapter and in Chapter 5. One technique is to run the digital device under test at its normal clock rate while monitoring for various control signals, such as reset, halt, memory read, flag, or clock. Questions such as ''Is there a clock signal?'' or ''Is there a memory read signal?'' are quickly resolved by noting if the probe indicator is flashing on and off, indicating that pulse train activity is present. For example, if there is no flashing indication when the probe is connected to the clock line, there is no clock pulse. Or if the probe indicator does not flash when connected to a memory read line in a microcomputer, the microprocessor is probably stuck somewhere in the program (since the memory read line is usually pulsed at each step of the program to open the RAM or ROM at each address).

Another useful technique is to replace the normal clock signal with a very slow clock signal from a pulse generator similar to the logic pulser described in Sec. 4-5. This allows changes in logic signals to occur at a rate slow enough that they can be observed on a *real-time* basis. (The basic single-stepping technique is discussed in Sec. 4-9.) Real-time analysis, coupled with the ability to inject logic-level pulses anywhere in the digital circuit with a logic pulser, and the means to detect logic-state changes with the logic probe, contributes to rapid troubleshooting and fault finding in the control lines. Unfortunately, logic probes are not as valuable with trying to locate faults in multiple-line circuits, such as the data and address buses.

4-5. LOGIC PULSER

Figure 4-3 shows a typical logic pulser being used with a current tracer (Sec. 4-6). The logic pulser is an in-circuit stimulus device that automatically outputs pulses of the required logic polarity, amplitude, current, and width to drive nodes low and high. A typical pulser also has several pulse burst and stream modes available. In operation, pulsers are hand-held logic generators used for injecting controlled pulses into digital logic circuits. As shown, the electronics are housed in a hand-held probe. Automatic pulse control is provided for TTL and many other logic families. Thus, the logic pulses are compatible with most digital devices. Pulse amplitude depends on the logic supply voltage (3 to 18 V), which is also the supply voltage for the logic pulser.

Pulse current and pulse width depend on the load being pulsed. The fre-

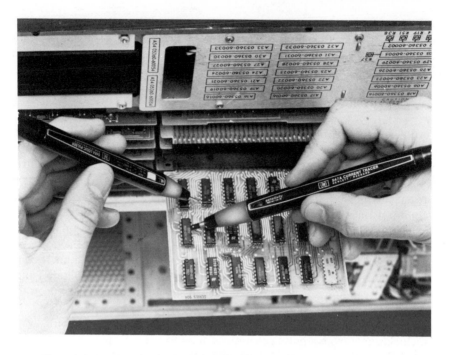

Figure 4-3   Hewlett-Packard 546A logic pulser (left) shown with Hewlett-Packard 547A current tracer. (Courtesy Hewlett-Packard Company.)

quency and number of pulses generated by the pulser are controlled by operation of a push-slide switch on the pulser probe. A flashing LED indicator located on the tip indicates the output mode.

The logic pulser generates pulses as programmed by the fingertip push-slide switch. The pulser is programmed by pushing the switch once for each single-pulse output, or a specific number of times for continuous pulse streams or pulse bursts at selected frequencies. The pulses are also applied to an LED indicator on the pulser tip. (The pulses applied to the LED are slowed down for visibility.)

Pressing the switch automatically drives an IC output or input from low to high (0 to 1) or from high to low (1 to 0). The high source and sink current capability of the pulser can override IC output points, originally in either the 0 or 1 state. The nominal 10-μs pulse width is long enough for even slow CMOS circuits to accept, but heavy circuit loads (such as TTL drivers) result in narrower pulses that limit the amount of energy delivered to the device under test.

The pulser output is three-state. In the off state, the pulser's high output impedance ensures that circuit operation is unaffected by probing until the pulser switch is pressed. Pulses can be injected while the circuit is operating, and no disconnects are needed. Some probes have a multipin kit or accessory that provides up to four pulses simultaneously (say to eliminate 4 bits of an 8-bit word on a data or address bus). To sum up, the logic pulser forces overriding pulses into logic nodes, and can be programmed to output single pulses, pulse streams, or bursts. The pulser can be used to force chips to be enabled or clocked. Also, logic circuit inputs can be pulsed while observing the effects of their outputs on the circuit under test.

4-5.1 Using the Logic Pulser in Digital Circuits

The following paragraphs provide brief descriptions of how the logic pulser can be used for troubleshooting. More detailed procedures and approaches are given in the remaining sections of this chapter and in Chapter 5. Logic pulsers are generally used with other instruments, such as with the logic probe (Sec. 4-4), current tracer (Sec. 4-6), logic clip (Sec. 4-7), and logic comparator (Sec. 4-8), for routine digital electronic troubleshooting.

4-5.2 Logic Gate Testing

A logic gate may be tested by pulsing the gate's input while monitoring the output with a logic probe, as shown in Fig. 4-4. The logic pulser generates a pulse opposite to the state of the input line and can change the gate output state.

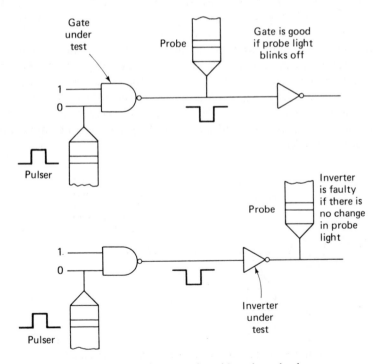

Figure 4-4 Logic gate testing with probe and pulser.

This assumes that the output of the gate is not clamped in its state by another input (such as a high on the other input of an OR gate).

If the pulse is not detected at the output, pulse the output line as shown in Fig. 4–5. If the output is not shorted to V_{CC} or common, the logic probe should indicate a pulse opposite to its original indication. If not, check for external shorts (solder bridges and the like) before removing the IC. Shorts at the inputs and outputs are best located by means of the current tracer described in Sec. 4–6.

4-5.3 Capacitor Pulse Generator

When a logic pulser is not available, some technicians use bus voltages to trigger circuits manually. That is, the pulse trains normally present on a particular circuit line or at a particular input are removed and replaced by pulses manually injected one at a time by momentarily connecting the line or input to a bus of appropriate voltage and polarity. This is not recommended, for two reasons: (1) some circuits can be damaged by prolonged application of the voltage; and (2) in other circuits, no damage occurs, but the results are inconclusive. For example, many faulty circuits operate normally when a bus voltage is applied (even momentarily) but do not respond to a pulse.

A better method is to use a capacitor as a pulse generator. The basic technique is shown in Fig. 4–6. Simply charge the capacitor by connecting it between ground and a logic bus (typically 5 or 6 V). Then connect the charged capacitor between ground and the input to be tested. The capacitor will discharge, creating an input pulse. Be sure to charge the capacitor to the correct voltage level and polarity. Generally, the capacitor value is not critical. Try a 1-μF capacitor as a starting value. Often, it is convenient to connect a lead of the capacitor to a ground clip, with the other lead connected to a test prod. The capacitor can be clipped to ground, and then the prod tip can be moved from bus to input, as needed.

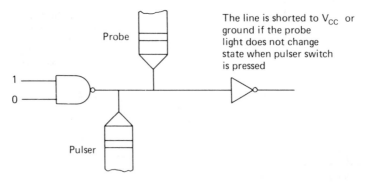

The line is shorted to V_{CC} or ground if the probe light does not change state when pulser switch is pressed

Figure 4-5   Testing for output shorts with pulser and probe.

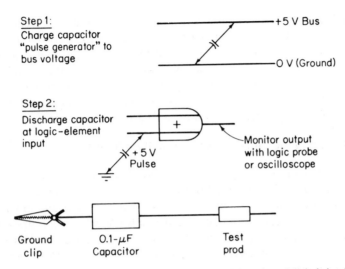

Step 1:
Charge capacitor
"pulse generator" to
bus voltage

+5 V Bus

O V (Ground)

Step 2:
Discharge capacitor
at logic-element
input

+

+5 V
Pulse

Monitor output
with logic probe
or oscilloscope

Ground
clip

0.1-μF
Capacitor

Test
prod

Figure 4-6   Capacitor pulse generator for single-pulse testing of digital circuits.

4-6. CURRENT TRACER

Figure 4–3 shows a current tracer being used with a logic pulser (Sec. 4–5). Such current tracers detect current activity on logic nodes by means of an inductive pickup at the probe tip. By adjusting the sensitivity control and observing the intensity of the tracer's lamp when placed on a pulsating line, you can identify current paths and relative magnitudes, and locate a bad device on a node. The advantages of the current tracer for troubleshooting are less obvious than those of the probe or pulser. This is because the current tracer deals with currents rather than the voltages found in nondigital troubleshooting. However, there is a great advantage once you gain confidence with the current tracer. Frustrating troubleshooting experiences with problems such as stuck nodes and shorted power supply lines can be dealt with directly. Without a current tracer, these types of problems may lead you to find the fault by cutting printed-circuit traces, snipping pins, or worse.

Current tracers are hand-held probes that enable the precise localization of low-impedance faults in many electrical systems (including typical digital electronic printed-circuit wiring). The current tracer senses the magnetic field generated by a pulsing current internal to the circuit or by current pulses supplied by an external stimulus such as a logic pulser. Indications of the presence of current pulses is provided by the lighting of the indicator lamp near the current tracer tip. Adjustment of current tracer sensitivity over the range 1 mA to 1 A is provided by a sensitivity control near the indicator. The current tracer is self-contained and requires less than 75 mA at 4.5 to 18 V from any convenient source. To sum up, the current tracer is used to track down the cause of stuck nodes, and can tell approximately how much pulse current is present and what path the current takes. When a logic pulser is used to inject current into a node without pulse activity, the impedance and the general nature of the problem (gate output, hard shorts, etc.) can be estimated. Then, the actual low impedance point can be found by tracing the path of the current from the logic pulser to that point on the node. The current either goes someplace it should not (a shorted line), or it enters a component that is stuck, shorted, or turned on.

4-6.1 Using the Current Tracer in Digital Circuits

The following paragraphs provide brief descriptions of how the current tracer can be used for troubleshooting. More detailed procedures and approaches are given in the remaining sections of this chapter and in Chapter 5.

The current tracer operates on the principle that whatever is driving a low-impedance fault node or point must be delivering the majority of the current. Tracing the path of this current leads directly to the fault. Problems that are compatible with this method are:

1. Shorted inputs of ICs
2. Solder bridges on printed-circuit boards
3. Shorted conductors in cables
4. Shorts in voltage distribution networks (such as V_{CC}-to-ground shorts)
5. Stuck data or address buses, including three-state buses
6. Stuck wired-AND structures

4-6.2 Basic Current Tracer Operation

Use of the current tracer is indicated when conventional troubleshooting reveals a low-impedance fault (such as a short). In typical operation, you align the mark on the current tracer tip along the length of the printed-circuit trace at the driver, and adjust the sensitivity control until the indicator lamp just lights. You then move the current tracer along the trace (line) or place the current tracer tip directly on the terminal points (IC pins) while observing the indicator light. This method of following the path of the current leads directly to the fault responsible for the abnormal current flow. If the driving point does not provide pulse stimulation, the terminal may be driven externally by using a logic pulser at the driving point. The following paragraphs describe troubleshooting techniques for some of the more common problems.

4-6.3 Wired-AND Problems

One of the most difficult problems in troubleshooting ICs of any type is a stuck wired-AND circuit. Typically, one of the open-collector gates connected in the wired-AND mode may still continue sinking current after the gate has been turned off. The current tracer provides an easy method of identifying the faulty gate. (Of course, if the gate is located in an IC, the entire IC must be replaced. However, you have still located the problem and can take whatever action is best suited to the situation.)

Referring to Fig. 4–7, place the current tracer on the gate side of the pull-up resistor. Align the mark on the tracer tip along the length of the printed-circuit trace (line) and adjust the tracer's sensitivity control until the indicator is just fully lighted. If the indicator does not light, use a logic pulser to excite the line as shown. Place the tracer tip on the output pin of each gate; only the faulty gate (or pin) will cause the indicator to light.

4-6.4 Gate-to-Gate Faults

When a low-impedance fault (short, full, or partial) exists between two gates, the current tracer and logic pulser can be combined to pinpoint the defect quickly. In Fig. 4–8, the output of gate A is shorted to ground. Place the pulser midway between the two gates, and place the current tracer tip on the pulser pin as shown. Pulse the line and adjust the current tracer sensitivity control until the indicator just lights. First place the current tracer tip next to gate A, and then gate B, while continuing to pulse the line. The tracer will light only on the gate A side, since gate A (the defect in this example) is sinking the majority of the current. If the tracer does not light when placed between the pulser and gate A, look for a short on the line between the pulser and gate B.

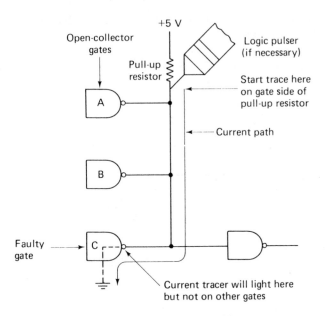

Figure 4-7 Locating wired-AND problems with a current tracer.

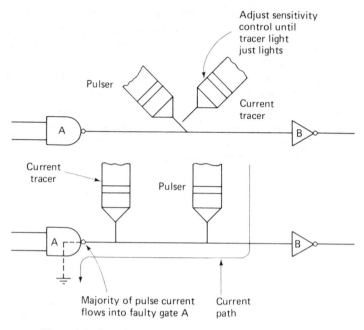

Figure 4-8   Locating gate-to-gate faults with a current tracer.

4-6.5 Solder Bridge/Cable Problems

When checking printed-circuit tracers (lines) that may be shorted by solder bridges or other means, start the current tracer at the driver and follow the trace. Figure 4-9 shows an example of an incorrect path due to a solder bridge. As the tracer probe follows the trace from gate A toward gate B, the in-

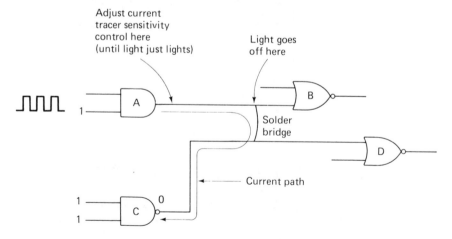

Figure 4-9   Locating solder bridges in printed-circuit lines with a current tracer.

dicator remains lighted until the tracer passes the bridge. This is an indication that current has found some path other than the trace. Visually inspect this area for solder splashes and the like. These principles also apply when troubleshooting shorted cable assemblies.

4-6.6 Multiple Gate Inputs

Another type of IC arrangement is the one-output, multiple-input configuration. Figure 4-10 shows this type of circuit being pulsed by a signal on the input of gate A. In this case, place the current tracer tip on the output pin of gate A and adjust the sensitivity control until the indicator light just comes on. Then check the input pins of gates B through E. If one of the input pins is shorted, that pin will be the only one to light the indicator.

Should the tracer fail to light when placed next to the output of gate A, it is a good indication that the problem exists in gate A. To be sure that this is true, use the current tracer as described for gate-to-gate faults (Sec. 4-6.4). If the circuit has no input signal, use a logic pulser.

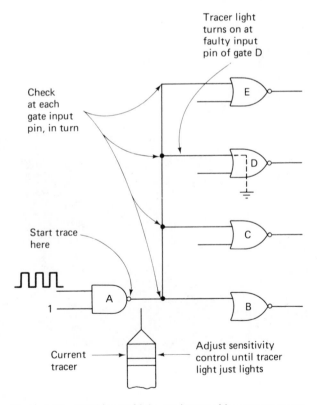

Figure 4-10 Checking multiple gate inputs with a current tracer.

4-7. LOGIC CLIP

Figure 4–11 shows a typical logic clip. Such clips are designed for logic-level determination on ICs using TTL and DTL circuits. Generally, clips can test flip-flops, gates, counters, buffers, adders, shift registers, and the like, but will not test ICs with nonstandard input levels or expandable gates. Clips will instantly and continuously show the logic levels (0 to 1) at all pins of a dual-in-line IC. All 16 input pins are electrically buffered to minimize loading on any circuit being tested.

Sixteen LEDs are the high and low (0 or 1) logic-level indicators. No power-supply connections need be made since the clip powers itself from the circuit under test by automatically locating the V_{CC} and ground pins of the IC. In use, you squeeze the thick end of the clip to spread the contacts, and place the clip on the IC to be tested. The LEDs on top of the clip indicate the logic levels at each connected IC pin. The clip may be turned in either direction. Each of the clip's 16 LEDs independently follows level changes at the associated input. A lighted LED corresponds to logic 1.

The real value of the logic clip is the ease of use. The clip has no controls to set, needs no power connections, and requires practically no explanation as to how it is used. Since the clip has its own gating logic for locating the ground

Figure 4-11 Hewlett-Packard 548A logic clip. (Courtesy Hewlett-Packard Company.)

and 5-V pins, the clip works equally well upside down or right side up. Buffered inputs ensure that the circuit under test will not be loaded down. Simply clipping the unit onto a TTL or DTL IC package of any type makes all logic states visible at a glance.

The logic clip is much easier to use than either an oscilloscope or a voltmeter when you are interested in whether a lead is in the 1 or 0 state, rather than the lead's actual voltage. The clip, in effect, is 16 binary voltmeters, and you do not have to shift your eyes away from the circuit to make the readings. The fact that a lighted LED corresponds to a logic 1 greatly simplifies the troubleshooting procedure. You are free to concentrate your attention on the circuits rather than on measurement techniques.

When the clip is used on a real-time basis (when the clock is slowed to about 1 Hz or is manually triggered) timing relationships become especially apparent. The malfunctions of gates, FFs, counters, and adders then become readily visible as all the inputs and outputs of an IC are seen in perspective. When pulses are involved, the logic clip is best used with the logic probe. Timing pulses can be observed on the probe, while the associated logic-state changes can be observed on the clip.

Figure 4-12 is the block diagram of a typical logic clip. As shown, each pin of the logic clip is internally connected to a decision gate network, a threshold detector, and a driver amplifier connected to an LED. Figure 4-13 shows the decision sequence of the decision gate network. In brief, the decision gate networks do the following:

1. Find the IC V_{CC} pin (power voltage) and connect it to the clip power voltage bus. This also activates an LED.
2. Find all logic-high pins and activate corresponding LEDs.
3. Find all open circuits and activate corresponding LEDs.
4. Find the IC ground pin, connect it to the clip ground bus, and blank the corresponding LED.

The threshold detector measures the input voltage. If the input voltage is not over the threshold level, the LED is not activated. An amplifier at the output of the threshold detector drives the LED. The LED indicates high (will glow) if the IC pin is above 2 V, and indicates low (no glow) if the pin is below 0.8 V. If the IC pin is open, the LED also shows a high (glow).

4-7.1 Using the Logic Clip in Digital Circuits

At present, logic clips will accommodate only 16-pin ICs. Thus, logic clips are of little value when checking digital devices such as microprocessors, where a typical microprocessor has 40 pins, and 24 pins is standard for most ROMs, RAMs, and I/Os. However, there may be 40- and 24-pin logic clips in the

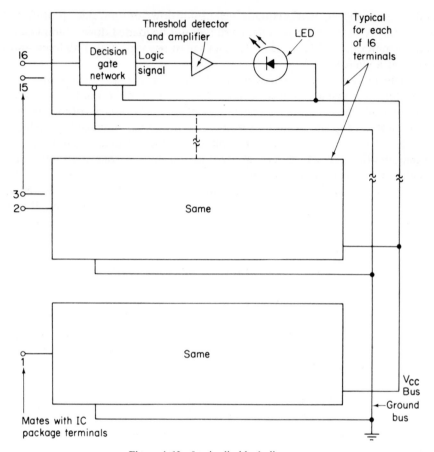

Figure 4–12   Logic clip block diagram.

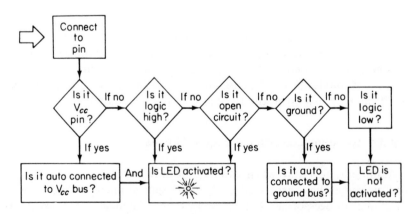

Figure 4–13   Logic clip decision sequence.

future. Also, in some microcomputers, there may be a 16-pin IC that has access to the data and address buses.

If a logic clip can have access to both buses, the clip can be used for a *program trace* function, using *single stepping,* as described in Sec. 4-9. In such a case, the clip serves to read out the data word at each address of the program. However, at this writing, the logic analyzer is the most practical instrument for making a rapid, thorough program trace.

4-8. LOGIC COMPARATOR

Figure 4-14 shows a typical logic comparator. Such comparators clip onto 16-pin ICs and, through a comparison scheme, instantly display any logic-state differences between the test IC and a reference IC. Logic differences are identified to the specific pin or pins of the IC with the comparator's display of 16 LEDs. A lighted LED corresponds to a logic difference.

In use, the IC to be tested is first identified. A reference board with a good IC of the same type is then inserted in the comparator. The comparator is

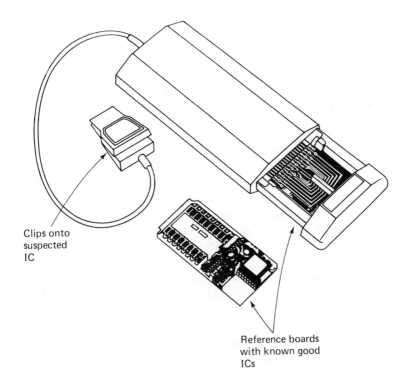

Clips onto
suspected
IC

Reference boards
with known good
ICs

Figure 4-14 Hewlett-Packard 10529A logic comparator. (Courtesy Hewlett-Packard Company.)

clipped onto the IC in question, and an immediate indication is given if the test IC operates differently from the reference IC. Even very brief dynamic errors are detected, stretched, and displayed.

The comparator operates by connecting the test and reference IC inputs in parallel. Thus, the reference IC sees the same signals that are inputs to the test IC. The outputs of the two ICs are compared, and any difference in outputs greater than 200 ns in duration indicates a failure. A failure on an input pin, such as an internal short, appears as a failure on the IC driving the failed IC. Thus, a failure indication actually pinpoints the malfunctioning pin.

As in the case of the logic clip, logic comparators are presently limited to 16-pin ICs. Thus, both instruments are of little value in present-day microcomputers. However, 40- and 24-pin comparators may be developed in the future.

4-9. LOGIC ANALYZER

Figure 4–15 shows a typical logic analyzer and some related displays. Logic analyzers are specialized instruments used for examining a number of logic signals (usually 16 or 32) during very specific portions of the long, complex data sequences present in microcomputer systems. Although logic analyzers are considered to be primarily microprocessor system development and system-level troubleshooting instruments, applications do exist in product service situations. Logic analyzers are useful for system software development because of their ability to trigger on and then follow specific program and timing sequences in the experimental stage. Logic analyzers can also provide a "map" of where the microprocessor system spends its time. These same capabilities can be used to analyze existing products. Good service documentation is generally required to describe proper setups, display outputs, and courses of action to be taken when incorrect program flow activity is observed.

Logic analyzers can be thought of as very specialized digital oscilloscopes that allow you to examine specific portions of the hardware and software sequences that occur in microprocessor systems. Logic analyzers can be especially useful as design aids for troubleshooting new product designs. The map-mode feature lends itself well to troubleshooting existing products. Logic analyzers require a greater understanding of hardware and software program flow than do the other digital troubleshooting equipment described in this chapter. Unlike this other equipment, however, logic analyzers need not be designed into the product to be useful (as is the case with the signature analyzer described in Sec. 4–10), and they provide much more detailed information about the system.

Figure 4–16 shows a logic analyzer display that represents address (column A) and data (column B) information in tabular form. This allows you to follow the sequence of events beginning with the instruction at address 06A2. Figure 4–17 shows the same sequence as it might appear in a typical program listing.

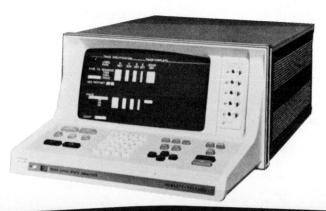

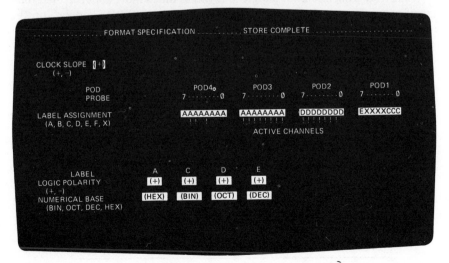

Figure 4–15 Hewlett-Packard 1610A logic analyzer and some related displays. (Courtesy Hewlett-Packard Company.)

Line number	A Hex	B Hex
000	06A2	36
001	06A3	00
002	0BF0	00
003	06A4	2C
004	06A5	C3
005	06A6	9B
006	06A7	06
007	069B	34
008	0BF1	03
009	0BF1	04
010	069C	7E
011	0BF1	04
012	069D	FE
013	069E	0A
014	069F	C2

Figure 4-16 Logic analyzer display representing address (column A) and data (column B) information in tabular form.

The 16 lines on the display of Fig. 4-16 represent a small portion of a long program sequence. The logic analyzer "captures" this sequence, stores the sequence in an internal memory, and can display the sequence indefinitely. The logic analyzer also allows you to look at activity before or after the operations shown. A logic analyzer is similar to an oscilloscope in many ways. Both instruments have signal inputs, a trigger timing circuitry, and a CRT display. The logic analyzer signal inputs differ from those of an oscilloscope in that the analyzer has 16 to 32 logic threshold-sensitive inputs that detect logic levels (1 and 0).

The triggering circuit of a logic analyzer is much more sophisticated than that of an oscilloscope. Analyzers can trigger on a particular address, on the Nth occurrence of that address (for software loops), after N clock cycles past

Line number	Operation	Comments
000, 1	MVI M, 00	Instruction
002	M ← 00	Execution M = location 0BF0
003	INR L	Instruction point to next M
004, 5, 6	JMP 069B	Instruction
007	INR M	Instruction
008	μP ← M	Execution M = location 0BF1
009	M ← M + 1	Execution
010	MOV A, M	Instruction
011	A ← M	Execution M = location 0BF1
012, 013	CPI 0A	Instruction
014	JNZ	Instruction

Figure 4-17 Data sequence as it might appear in a typical program listing.

that address (for delay), or after a specified sequence of trigger addresses (to detect a particular program path). The trigger words need not be addresses, but can be data, control, or any combination of logic signals.

Like oscilloscopes, logic analyzers can display the data that occur immediately after the trigger event. Unlike most oscilloscopes, logical analyzers can also display data prior to the trigger event. This capability, called *negative time* recording, can be used for troubleshooting by choosing a faulty system operation to be the trigger event and then observing the events that led up to the event.

Logic analyzers use the clock of the circuit they are connected to for timing. A new line of display information is generated for each clock input. Whereas oscilloscopes are referred to as *time domain* instruments, logic analyzers are often called *data domain* instruments. In an oscilloscope, time advances from left to right across the display, using an internal time base. In a logic analyzer, time advances from top to bottom, one line per clock input. The display of a logic analyzer is a tabular listing of digital data. In Fig. 4–16, the display is formatted in hex numerals, but can also be displayed in binary, octal, or decimal.

4-9.1 The Need for a Logic Analyzer in Troubleshooting

Before going into how logic analyzers are used with programmed systems, let us discuss the need for a logic analyzer (also known as a logic-state analyzer). The classic approach for troubleshooting any programmed device is to monitor a significant system function (such as the data and address buses), go through each step in the program, and compare the results with the program listing for each address and step. This is sometimes known as a *program trace* or, more simply, a *trace* function. One technique for this procedure is called *single stepping*. With the single-stepping approach, you remove the normal clock pulses and replace them with single one-at-a-time pulses obtained from a switch or pushbutton. This permits you to examine and compare the data at each address with those shown in the program.

For example, assume that each of eight lines on a data bus is connected to a multitrace oscilloscope so that the bit on each line appears as a pulse on a corresponding trace. A pulse on the oscilloscope trace indicates a binary 1, whereas the absence of a pulse indicates a binary 0 on that line. A simplified version of such a test setup is shown in Fig. 4–18.

The test is started by applying a single pulse to the reset or clear line. Then sufficient single pulses are applied to the clock line until address number 1 (0001) appears on the address bus. Now assume that the program listing shows that a hex 7F (binary 0111–1111) data byte should be on the data bus at address 0001, but that the oscilloscope traces show binary 0011–1111. Obviously, the wrong instruction will be applied to the microprocessor and the system will malfunction. This could be caused by a broken line on the data bus, by a defect

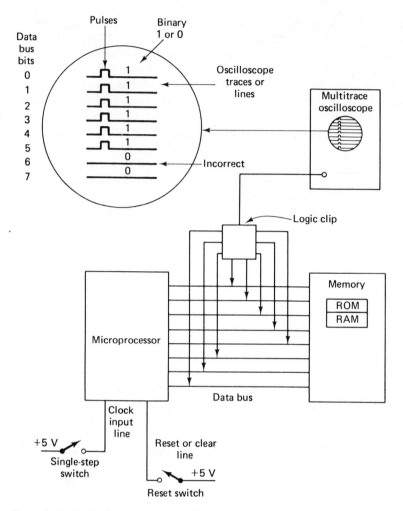

Figure 4–18 Basic microcomputer troubleshooting approach using multitrace oscilloscope and single-stepping.

in the memory, by the absence of a memory-read pulse or a memory-read pulse that appears at the wrong time (opens the memory data buffer too soon or too late), or by several other possible causes. However, you have isolated the problem and determined where it occurs in the program.

4-9.2 Timing Problems

If the problem appears to be one of timing, the oscilloscope can be used to check the time relationship of the related pulses. For example, the oscilloscope can be connected to the data bus, address bus, and read line, as shown in Fig. 4–19. The oscilloscope then shows the time relationships among the pulses on these lines.

In this very simplified example, the read pulse must hold the memory data buffer closed until the selected address pulses appear on the address bus (sometimes known as the *valid address* point), must hold the buffer open just long enough for all 8 data bits to appear on the data bus, and then must close the buffer until the next address is applied. In a practical case, an entire timing diagram can be duplicated on a multitrace oscilloscope.

4-9.3 Logic Analyzer Displays

While single stepping and a check of system timing can pinpoint many digital equipment problems, there is an obvious drawback. Typically, a data byte is 8 bits and thus requires eight clock pulses or eight one-at-a-time pushes

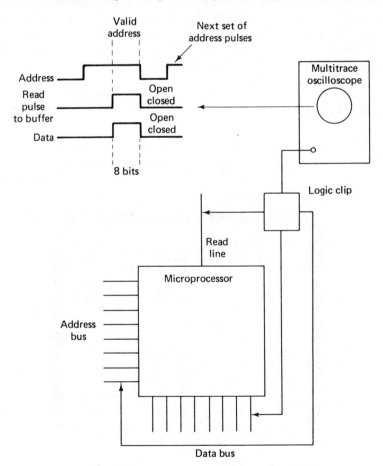

Figure 4-19   Basic microprocessor troubleshooting approach for timing problems.

of the single-step button. Since all program steps require at least one byte (and often two or three bytes, possibly 24 bits), you must push that button many times if the malfunction occurs at step 0333 of the program. This means that you must spend endless hours comparing program listings against binary readouts at addresses. (If you are already familiar with the troubleshooting of programmed devices, you know that the most time-consuming part of the task is in making such comparisons.)

The logic analyzer overcomes this basic problem by permitting you to select for display the data at a particular address. The logic analyzer then runs through the program at near-normal speed (typically a fraction of a second) and displays the selected data between the desirable *breakpoints* (points in the program before and after the area of interest). Figure 4–15 shows the related displays of a typical logic analyzer. A logic analyzer is essentially a multitrace oscilloscope combined with electronic circuits to produce special displays. The electronic circuits that produce these displays are sometimes known as *formatters*. The logic analyzer shown in Fig. 4–15 is operated by a keyboard. Other logic analyzers use switches and controls. No matter what control system is used, the logic analyzer is capable of three basic displays (timing, tabular, and map).

4-9.4 Timing Display

The multitrace feature of the logic analyzer can be used to reproduce *timing diagrams* in a manner similar to that of a conventional multitrace oscilloscope. Figure 4–20 shows a typical timing diagram display as produced

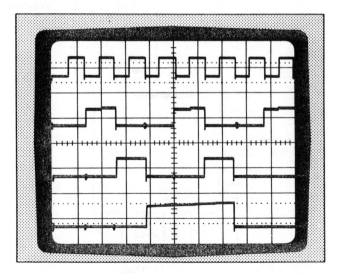

Figure 4–20   Typical timing diagram (time-domain) display as produced on a logic analyzer.

on a logic analyzer. Some logic analyzers have a feature called *timing analysis*. The display is similar to that of an oscilloscope timing diagram, except that the display is "digitized" (or rounded off) to discrete time and voltage limits to give a clear graphic display without specific waveshape information. Figure 4-21 is a typical timing analysis display. With such a display, glitches (narrow pulses) can be detected by means of internal detection and stretching circuits. Many inputs can be observed at once. By taking full advantage of the triggering features, you can observe the timing of a small, specific portion of a long program sequence. The elaborate triggering is made possible by the fact that waveforms are stored in digital form in a RAM within the logic analyzer.

4-9.5 Tabular Data Display

The data display format of the logic analyzer (sometimes called the *data domain* format or display) is used to display data bytes (as they appear on the data and/or address buses) in binary or hex (or octal) form. In effect, the bytes appear as they would on paper (1s and 0s rather than the presence or absence of pulses). Figure 4-22 shows that several data bytes can be displayed (in binary) simultaneously. This makes it possible to check the data words before and after the selected data area of interest in the program. With some logic analyzers, the selected data word is indicated by extra brightness of the display. In Fig. 4-22, the top data word is the selected word (the breakpoint starts at this word and continues for 16 steps or words).

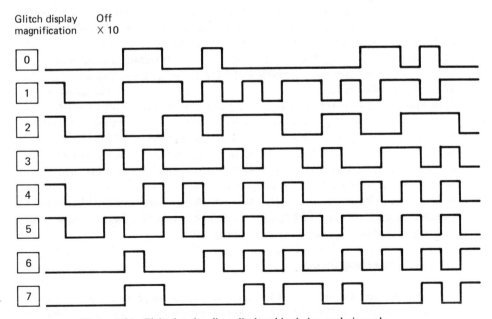

Glitch display   Off
magnification   X 10

Figure 4-21   Eight data bus lines displayed in timing analysis mode.

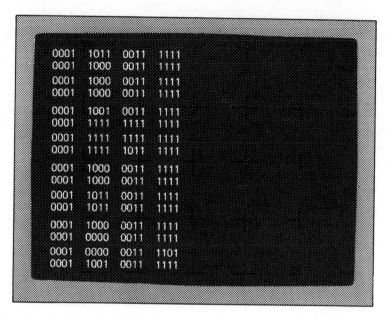

Figure 4-22 Typical tabular data (data-domain) display as produced on a logic analyzer.

To use the tabular data format, you connect the logic analyzer to the data and/or address buses by means of patch cords or probes supplied with the analyzer. Then you select a particular data word and/or address breakpoint from the program listing by means of the logic analyzer controls (keyboard or switches) and start the program. The digital system then runs through the complete program, but only the desired portion of the program is displayed.

Figure 4-23 shows a comparison of a typical program listing versus a logic analyzer display (in binary). In this example, note that both the address and data bytes are shown for 16 steps of the program. However, only the first nine addresses are of interest, with address 0004 being of special interest. Compare this tabular display with that of the timing display (Fig. 4-20). Even those readers not familiar with troubleshooting will quickly realize the advantages of a tabular display for digital troubleshooting and for debugging of programmed devices. Keep in mind that the program can be operated at near its normal speed, and can be examined line by line, 16 lines at a time. The next 16 lines can be selected by the simple setting of a switch or touch of a key. Compare this with single stepping through the entire program!

4-9.6 Mapping Display

Another display unique to logic analyzers is the mapping display or mode. A mapping display is formed by connecting the most significant bits (MSB) of a data word to the vertical deflection circuits of the logic analyzer (those circuits

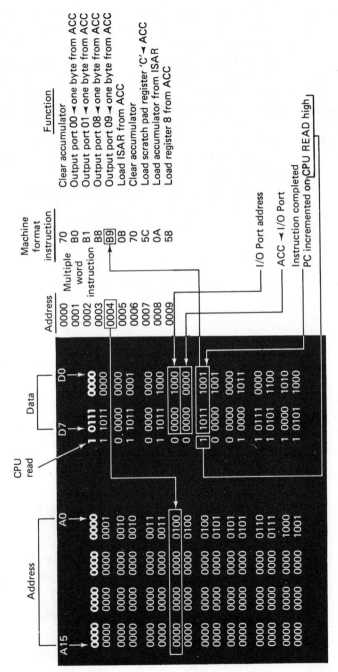

Figure 4-23 Comparison of a typical program listing versus a logic analyzer tabular data display.

189

that cause vertical deflection of the oscilloscope trace), while the least significant bits (LSB) are connected to the horizontal deflection circuits. This produces a series of dots as shown in Fig. 4–24. In the mapping mode, the display is
an array of 256 dots instead of a table of 1s and 0s. In effect, the analyzer CRT
is used as an X-Y grid of up to 256×256 (65,636 or 2^{16}) points. Each dot
represents one possible combination of the 16 input lines, so that any input is
represented by an illuminated dot. An input of all 0s (0000 in hex) is at the upper
left corner of the display, whereas an input of all 1s (FFFF in hex) is at the lower
right. The dots are interconnected so that the sequence of the data changes can
be observed. The interconnecting line gets brighter as it moves toward a new
point, thereby showing the direction of data flow.

  When the mapping display is used to monitor a digital equipment pro-

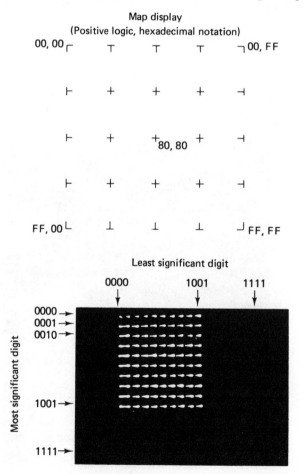

Figure 4–24 Typical mapping display as produced on a logic analyzer.

gram or a portion of a program, the display assumes a unique pattern (sometimes called a "map signature"). Once you have learned to recognize the patterns for various programs, it is relatively easy to tell at a glance if the program is proceeding normally. Compare the table and map display of Fig. 4-25. The same data bytes (16 words, 16 bits per word) are in tabular form (Fig. 4-22 top) and plotted in a map format in Fig. 4-22 (bottom).

Some logic analyzers are provided with a *cursor* function to help locate specific points in the mapping display. Generally, the cursor is a bright circle

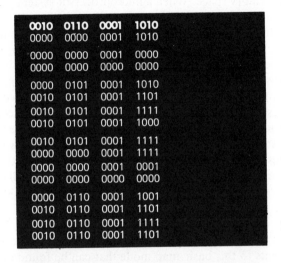

0010	0110	0001	1010
0000	0000	0001	1010
0000	0000	0001	0000
0000	0000	0000	0000
0000	0101	0001	1010
0010	0101	0001	1101
0010	0101	0001	1111
0010	0101	0001	1000
0010	0101	0001	1111
0000	0000	0001	1111
0000	0000	0001	0001
0000	0000	0000	0000
0000	0110	0001	1001
0010	0110	0001	1101
0010	0110	0001	1111
0010	0110	0001	1101

Figure 4-25   Comparison of a tabular display and mapping display as produced on a logic analyzer.

that can be manipulated over the display by the analyzer controls. In use, the cursor is positioned over an area of interest on the map, and the address or data word is read out on the analyzer controls. For example, as shown in Fig. 4-26, there is a gap between the sixth and seventh lines of display. By positioning the cursor over the last dot in the sixth line and reading the corresponding word on the analyzer controls, you know the exact address or data word at which the malfunction occurs.

In a typical mapping display, one point can be used to represent each address in a 64K memory space. When a microprocessor system is running a program, one of the 64K points on the CRT is lit each time the microprocessor specifies an address. The more often the microprocessor points to that address, the brighter the point becomes. As the program executes sequential instructions, successive points are illuminated. When jumps occur in the program, they can be observed on the CRT as a jump from one cluster of points to another. To improve resolution, the display can be expanded to magnify a region of interest. Figure 4-27 shows an 8085 microprocessor loop displayed in the expanded mode. The loop begins at address 0800. When the loop gets to address 0808, it stores a data byte at address 3000. At address 0810, the loop jumps to address 0F20 to continue the program. Address 0F20 contains an instruction that causes a data byte to be stored at address 3838. Finally, a jump instruction at 0F24 causes the program to go back to address 0800 and repeat the loop.

The map display shows where the system is spending its time (which subroutine and peripheral addresses it is using). There are several variations and enhancements to this basic map mode that can be found on specific logic analyzers. The map function is particularly useful for checking a microprocessor system that shows bus activity (by means of a logic probe) but is not functioning properly. The map of a good product can also be compared with that of a bad one as an aid in identifying possible problem areas.

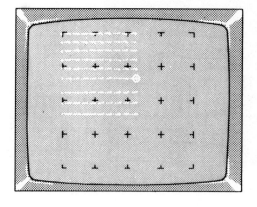

Figure 4-26   Using the cursor function of a logic analyzer mapping display to locate a malfunction in a program.

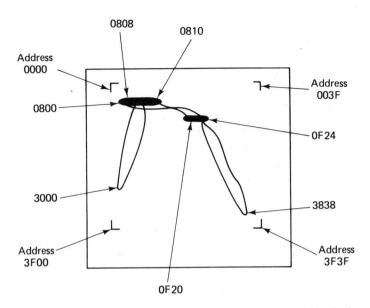

Figure 4–27   An 8085 microprocessor loop displayed on a logic analyzer in the expanded mapping mode.

4–10. SIGNATURE ANALYZER

Figure 4–28 shows a typical signature analyzer. Signature analysis (SA) is an easy-to-use and highly accurate technique for identifying faulty logic circuits. The signature analyzer can convert the long, complex serial data streams present in microprocessor system logic circuits into four-digit "signatures." These signatures tell whether or not the circuit (or a particular point or "node" in the circuit) is acting properly.

The character set used for signature analysis contains 16 characters (0–9, A, C, F, H, P, U). By using this character set, the confusion resulting from the seven-segment representation of the hex characters "b" and "6" is eliminated. All the characters are also readily distinguished from each other. In use, points in the suspected circuit are probed by the signature analyzer until a signature is found that does not agree with the one documented in the service manual. The signal path is then traced back until a correct signature is found, localizing the fault. Once the faulty point is found, the bad component at that point usually becomes apparent (or can be located using the current tracer, logic pulser, and probe, etc.). Signatures at points in digital circuits are used in much the same manner as the voltage and waveform information found on analog circuit service schematics (as discussed in Chapter 1).

Signatures are meaningful because they are generated by a test stimulus

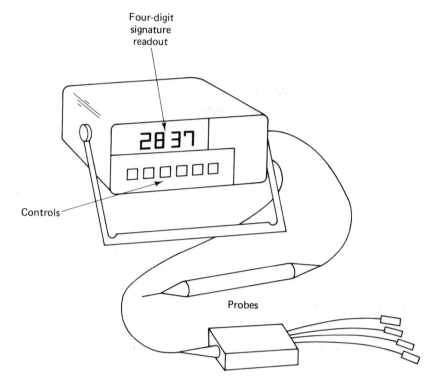

Four-digit
signature
readout

Controls

Probes

Figure 4-28 Typical signature analyzer.

program, provided by either the product under test or an external adapter. This stimulus program exercises specific portions of the circuit in a controlled, repeatable manner. You may wonder how proper signatures are determined by the manufacturer. The normal procedure is to take a known-good product and gather signatures from it, point by point. This approach is taken because it is very difficult to predict the tremendous amount of complex information contained in each signature.

Signature gathering is a time-consuming process, and errors can and do occur. Therefore, it is essential that you check the product serial number against the service manual when using signature analysis. Also look for any product errata sheets in the service literature. If even a single word of ROM is changed because of a product revision, most or even all of the signatures may need to be changed in the service literature. There is no such thing as a signature being almost right; the signature is either right or wrong. The signature 8F37 is no more related to 8F38 than to CCCC. If you must troubleshoot the same type and model of equipment on a regular basis, and there is no signature analysis information available in the service literature, you might do well to gather a set of valid signatures on equipment known to be in *perfect operating condition*. Normally, you note the signature at each pin of each IC. Make absolutely certain that operation is normal; make note of any changes in signatures for different modes of operation (just as you would for the waveform, voltage, and resistance measurements described in Chapter 1).

The signals required by the signature analyzer to generate a proper signature are data, start, stop, and clock. The data input receives data from the point under test. The start signal, provided by the product under test, tells the signature analyzer when to begin looking at the data, and the stop signal tells the analyzer when to stop. Between the start and stop signals, data are processed every time a new clock input occurs. The connection points for these inputs must be specified in the service literature.

Figure 4-29 shows the signatures for an IC as they might appear in signature tables of a service manual for digital equipment. In most cases, a signature is listed for each pin. However, you may find a great variety of ways in which signatures are listed (up to and including no listing whatsoever!). In some cases, signatures are shown on the schematic or logic diagrams. In the case of the system shown in Fig. 4-29, if the pin is tied directly to ground or +5 V, the table entry simply indicates ground (GND) or V_{CC} rather than giving the signature. The V_{CC} signature (typically 0001) is listed at the beginning of the table, and the GND signature is usually 0000. An X usually indicates that the pin is unused, and any signature found at that pin is meaningless. The table may also show a 1 or 0 signature, which means the same thing as V_{CC} or ground ex-

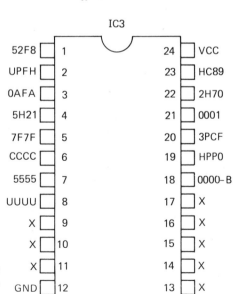

V_{cc} signature: 0001

Figure 4-29 Signatures for an IC as they might appear in signature tables of a service manual for digital equipment.

cept that the signal is gate output, not tied directly to V_{CC} or ground. Finally, the table may show a V_{CC} or ground signature with a B after the signature (such as on pin 18 in Fig. 4–29). This means the same thing as a 1 or 0 except that the light on the signature analyzer's probe tip should be blinking. The B means that although the signal is always at the same logic level when the clock edge arrives, at other times the signal is at a different level.

5

Microprocessor-Based
Digital
Troubleshooting

It is assumed that you are familiar with the basics of digital circuits and troubleshooting, at least at a level found in the author's best-selling *Handbook of Digital Electronics* (Englewood Cliffs, N.J.: Prentice-Hall, Inc., 1981). The information in this chapter is based on the troubleshooting methods covered in that book, and in Chapter 1 of this book. However, this chapter is devoted primarily to the problems associated with troubleshooting microprocessor-based digital equipment. Microprocessor systems can be thought of as an extension of traditional digital logic. Many of the components, circuit designs, and troubleshooting tools and techniques are the same. However, there are some differences. Microprocessor systems are bus-structured, and many of the devices on the bus are complex LSI devices. The signal activity between the devices on the buses is constant and complex. It is often useful to break the data bus, which is the system's main feedback path, to help isolate a fault that causes the entire system to malfunction.

Although basic digital troubleshooting techniques (such as troubleshooting trees) provide an orderly approach to locating system faults, they are not always adequate. There are numerous techniques, procedures, and tricks that can be effective in diagnosing, isolating, and locating faults in microprocessor-based products. Many of these techniques are discussed throughout this chapter. We start with a discussion concerning some troubleshooting problems unique to microprocessors. These procedures are based on the techniques recommended by the troubleshooting masters at Hewlett-Packard.

5-1. SOME TROUBLESHOOTING PROBLEMS
UNIQUE TO MICROPROCESSORS

One of the major problems in troubleshooting any microprocessor-based system is the difficulty in testing system functions as well as individual components. First, microprocessors operate at such high speeds, with an entire program being completed in seconds (or a fraction of a second), that it is impossible to monitor any function (system or component) in real time. Similarly, you cannot stop the system function partway through the program, say at the point where the program or system appears to fail, as you can with some non-microprocessor digital devices. You must make measurements while the microprocessor is running. Of course, you can single-step through a program (as described in Sec. 4–9), using a logic probe and pulser, but this is tedious, time consuming, and still does not provide a true picture of system operation. It is quite possible for a particular function to appear good when pulsed and measured under static conditions, but fail when the normal signal is applied during dynamic operating conditions.

The fact that microprocessor systems are bus-structured creates another problem in testing. Buses make it possible for many devices to be connected together on a single line. Unless you can remove each device, one at a time, it is difficult to tell which device is bad. The current tracers described in Sec. 4–6 are helpful in finding a bad device on a line connected to many devices. One particular problem with any bus-structured system is that the bus acts as a digital signal feedback path, and tends to propagate errors through good circuits and then back to the fault source. The best way to overcome this problem is to open the feedback path when possible. We discuss the procedures for opening the feedback path in this chapter.

Microprocessor-based systems are sequential in nature. That is, program flow depends on a long sequence of instructions and events. If even a single bit of information is incorrect, the whole system will fail. Noise glitches and bad memory bits are the most common sources of single-bit errors. For example, if a noise glitch is of the same approximate amplitude as the pulses on a data or address bus, the glitch can be confused with a logic 1. Similarly, if a logic-1 bit is stored in memory, and the bit is degraded (say the bit amplitude is below the normal threshold of other pulses), the bit can appear as a logic 0. If either of these conditions were to cause the system to go to a wrong address during any portion of a program, the remainder of the program would be incorrect. These failures are difficult to pinpoint because the entire system appears to be operating incorrectly, when actually only one bit is in error.

The devices connected to microprocessor buses are often complex (ROMs, RAMs, I/Os, etc.). Such devices cannot be tested using simple stimulus–response testing (apply a single input and look for a corresponding output). In a few cases, it is possible to verify operation of these devices if you can observe the function the devices are supposed to perform. This is not always practical. In some cases, the function is not clearly defined. In other cases, you simply cannot observe the function. The only course is to substitute a known-good device. Substitution sounds simple on paper, but can be a tedious job, especially where the device is an IC that must be unsoldered and soldered. The logic comparator discussed in Chapter 4 can be helpful when substitution is indicated as a logical step in troubleshooting. It is also possible to "piggyback" ICs (as described in this chapter). However, the problem is usually solved only by physical substitution.

Now that we have reviewed some of the general problems, let us discuss a few specifics in troubleshooting microprocessor-based systems.

5-1.1 Power-Up Reset Problems

Virtually all microprocessor systems have a power-up reset function. That is, a reset pulse is applied to the microprocessor when power is first applied, or under certain conditions, causing the program to go to zero (typically the program counter in the microprocessor will go to 0000 or possibly 0001). The program then starts when a certain condition occurs, or automatically, depending on the system.

If the reset pulse is absent, too short, not of the correct amplitude, noisy, or too slow in transition (a sloping pulse instead of one with steep edges), the program can start at the wrong point, resulting in out-of-sequence, partial, or no-reset functions. Problems can also occur in reset circuits that are susceptible to power-supply glitches. Even when Schmitt input circuits are used, slow edges on reset pulses can cause timing problems from one device to another within some systems. This can cause some of the devices to power-up before the others, resulting in erroneous behavior. A too rapid on–off–on system power sequence will fail to restart some systems, and it may be necessary to increase the off-time to allow the power supplies and restart circuits to discharge.

None of these reset failures necessarily prevent the system from running. Often, the system will run partway through the program and then stop, or lock up on a meaningless program loop, or even perform most of the normal operations. The point to remember is that the system must complete the power-up reset sequence to ensure that all the test, control, and initialization operations necessary to bring the system up have been performed. Fortunately, power-up circuits are normally operative only when the system is initially powered-up, and can be monitored at that time with storage oscilloscopes, logic analyzers, and (in some cases) signature analyzers. Also, it is possible to manually overdrive the reset circuits, and control the circuit externally during test. For example, you can apply a fixed d-c reset pulse and see if the system returns to program start.

5-1.2 Clock Problems

All microprocessor systems have some form of clock signal that keeps all system functions synchronized. In some systems, the clock is a single pulse that appears at a fixed repetition rate, and is applied to all ICs in the system via a clock line. Other systems have more than one clock pulse, possibly two pulses that appear on the clock line with some fixed delay or phase relationship between the pulses. Some microprocessors have a built-in clock, whereas others require an external clock. Clock speed can be controlled by quartz-crystal circuits, or can be timed by simple RC (resistor–capacitor) circuits.

No matter what clock system is used, total failure of the clock circuits is usually easy to find. The system does not run, and there are no clock pulses on the clock line. It is when the clock is bad (too fast, too slow, not of the correct amplitude or shape, etc.) that troubleshooting is difficult. This is because the system will run, but not properly. Therefore, the clock pulses should be one of the first functions checked in microprocessor troubleshooting (immediately after or at the same time as when the reset pulses are checked).

There are a number of malfunctions that can result from a bad clock. For example, clock problems can show up as a failure to function at all (no activity on data or address buses), the ability to function only open-loop (free-running), or when a meaningless and undefined program sequence occurs. Since these same symptoms can be caused by other malfunctions, your only course is to check for correct clock pulses (early in the troubleshooting sequence).

Many microprocessors are sensitive to clock speed (both too-slow speeds as well as too-fast speeds). Since many systems run at the maximum rated speed of the microprocessor, if clock speed is even slightly over the limit, the system can fail. Also, if the system runs too slow, dynamic storage cells on ICs in the system may fail. Both of these problems are more likely to occur when RC clock circuits are used instead of the more accurate and stable crystal-controlled circuits. However, crystals can sometimes break into their third overtone oscillation mode, causing a much higher than expected clock rate. Also, when the system requires multiphase and nonoverlapping clocks, the timing/phase requirements are usually very stringent. In some cases, clock voltage levels are not necessarily compatible with all system components, but may be much wider in voltage swing.

Microprocessor clock specifications can be found on the device data sheets (when you cannot find the clock data in the service literature for the system). The actual clock pulses in the system can be monitored on an oscilloscope (for amplitude, width, shape, etc.), and on a frequency counter (or possibly an oscilloscope) for correct speed.

5-1.3 Interrupt Problems

One function found in most microprocessor systems is the capability of responding to an interrupt signal or service request (say from an external video terminal, disk, or tape reader). An interrupt request causes the control logic to interrupt program execution temporarily, jump to a special routine to service the interrupting device, and then automatically return to the main program. When there are several interrupts, each interrupt is usually assigned a priority. This eliminates a conflict when two or more interrupts occur simultaneously.

Stuck or noisy interrupt lines can cause faulty system operation. The system may work with a stuck line but will do so very slowly (spending most of the time servicing the "phantom" interrupt). Noisy interrupt lines can cause sporadic system changes to occur, or peripheral inputs or outputs may take place at improper times. Sometimes the system will not respond at all to certain I/O (input/output) devices, which can occur when a higher-priority interrupt has disabled the lower ones.

Interrupt line activity can be monitored with a logic probe, logic analyzer, or oscilloscope. Interrupts are asynchronous in nature, and can often be manually controlled (enabled or disabled) for testing purposes. For example, in a typical video terminal, an interrupt occurs each time a key is struck.

5-1.4 Memory Problems

Most microprocessor systems contain ROMs (read-only memories) and RAMs (random-access memories). ROMs are generally used to store the program, routines, tables, and so on, and can be programmable or nonprogrammable. RAMs are generally used to manipulate the data bytes during the program, and can be considered as read/write memories (where data bytes can be written in and read out easily). Some microprocessor systems have self-test programs to check the ROMs and RAMs. Such self-tests often occur during the power-up reset sequence.

There are many ways in which a memory defect can cause a system failure. If there is a self-test program included in the system, the memory failure will be found quickly (unless the memory failure prevents the self-test from being completed). RAMs are generally harder to test than ROMs, and generally produce more disastrous results when a failure occurs. For example, when even a one-bit error occurs in that portion of a RAM reserved for the stack (an address where the program counter information is stored temporarily), the entire program "crashes." From a practical hardware standpoint, it is generally necessary to substitute a suspected RAM. If the RAM is dynamic rather than static (where external data refresh circuitry is required for the dynamic RAM), testing and troubleshooting are even further complicated since any failure in the refresh circuit can cause a good RAM to apparently fail.

Although ROMs fail as frequently as RAMs, ROM failures are generally not as disastrous, and are generally easier to locate. A typical ROM failure is where the program runs normally until it reaches a defective address in the ROM. (Of course, if this is at the beginning of the program, there will be total failure thereafter.) ROMs can be effectively tested during a power-up self-test (if such tests are designed in). ROMs can also be tested by other techniques if no self-test is provided. One such technique involves free-running the system and then using a signature analyzer either to verify documented signatures or compare the outputs of a suspected ROM with that of a ROM in a known-good system (if you are so fortunate as to have a known-good system!).

The programmability of microprocessor-based systems can be used to great advantage in system self-tests. Programs stored in the ROM can test ROMs, RAMs, and the microprocessor. In some cases it is also possible to test the I/O. Software can also be used to provide a stimulus for an external test instrument such as a signature analyzer.

ROM self-tests. The most common technique for testing ROMs during power-up reset is called a *checksum*. When the ROM is programmed, all the ROM words are added together, ignoring any carries that result. This number is complemented and stored in the last (or sometimes the first) word of the ROM, so that when all the words are added together (including the checksum stored in the last byte), the result is zero. If the total is not zero at the end of the test sequence, something is wrong with the ROM. (In actual practice, the checksum is usually calculated to make the total a specific number other than zero, and this number is compared with a corresponding number during the power-up sequence. If the numbers do not compare, the system is halted.)

Unfortunately, the checksum is not totally reliable. A checksum can detect any single-bit error, and most multiple-bit errors, but there are combinations of two or more errors that could provide the correct checksum. For example, if one error caused an undesired bit to be added, and another error caused a desired bit to be omitted, the checksum would still appear correct. Thus, the checksum is a negative test. That is, if the system fails the checksum test, something is wrong, probably (but not necessarily) the ROM. On the other hand, if the checksum is good, the ROM is probably good.

RAM self-tests. RAMs are tested by writing a pattern into the memory, reading the pattern back, and the verifying if the pattern has or has not changed. The *checkerboard* is one of the many different patterns that can be used for RAM self-tests. Many other patterns used to test RAMs are specifically aimed at detecting various failure mechanisms within the RAM. With the checkerboard, all the bits at each address in the RAM are set to alternating 1s and 0s. Once all memory locations have been tested, the pattern is repeated with each bit reversed, verifying that each bit of the RAM can store a 1 and a 0.

No memory test can guarantee complete accuracy, even though the test may show that each bit can store a 1 or a 0. Some RAMs are pattern sensitive. For example, one location might correctly store 01010101 and 10101010, but fail when 01111000 is stored. Even for a small RAM, it takes an extremely long time to test every possible pattern sequence. For this reason, RAM test credibility is generally much lower than that of ROMs. As with the checksum test, if a

RAM passes the system self-test program (such as a checkerboard), the RAM is probably good. If the system fails a checkerboard, something is wrong (probably the RAM).

5-1.5 Signal Degradation Problems

The long parallel bus and control lines present in medium-to-large microprocessor systems are sometimes susceptible to crosstalk and transmission line problems on critical lines (such as clocks and chip enables). These problems can show up as glitches on adjacent signal lines (crosstalk) or ringing (overdriven pulses) on the driving line (causing multiple transitions through a logic threshold). Either of these situations can inject faulty data or control signals that are very difficult to detect. This problem is most common when signal lines are long and already taxing the timing noise margins of the system. When extender cards are added to these systems or high-humidity conditions exist, failures may occur. Cross-coupling of lines on extender cards can be a problem when fast signal transition lines (such as Schottky gate outputs) run alongside other signal lines, even when they are on opposite sides of a PC board.

5-1.6 Troubleshooting Aids

Well-written service literature for microprocessor equipment includes theory of operation, schematics, block diagrams, and troubleshooting trees. The schematics often provide too much detailed information, making it difficult to see the "big picture." Similarly, the theory of operation may or may not relate closely to the hardware. Therefore, for many experienced troubleshooters, the block diagrams can supply the right amount of information to understand how the various parts of the circuit work together.

A troubleshooting tree is a graphic means of showing the sequence of tests performed on a product under test. These trees are often drawn as flowcharts in which the results of each test determine what step is taken next. The use of troubleshooting trees for repairing microprocessor-based products can save considerable time and effort.

Figure 5-1 shows a portion of the troubleshooting tree for the Hewlett-Packard HP3455A Digital Voltmeter. Theoretically, the tree should lead you to the fault by means of the actions taken and decisions made along the tree. Unfortunately, such is not always the case. A perfect troubleshooting tree must consider all possible failures, a difficult criterion for the person writing the troubleshooting tree to meet. Also, troubleshooting trees tend to be fairly generalized, lacking the specifics desired for making tests and decisions. Few troubleshooting trees provide practical information about how a specified test or measurement relates to what the circuit does or is supposed to do. If the troubleshooting tree fails to direct you to the actual fault, you may be left at a

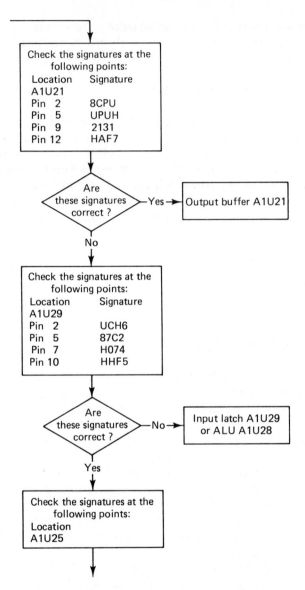

Figure 5-1   Portion of a typical troubleshooting tree.

dead end, with no idea of where to go next. However, the troubleshooting tree is often your best guide (at least as a starting point).

There are good and bad troubleshooting trees. The good ones seldom lead to a dead-end and provide a logical, well-directed sequence of tests and measurements, requiring a minimum level of understanding of the circuit under test. Often, even a poor troubleshooting tree includes advanced techniques such as signature analysis to simplify the procedure, and can thus save time and effort.

5-1.7 Interface and Multiplexed I/O Problems

Many microprocessor systems interface with other systems through external communications lines (IEEE-488, RS-232C, telephone modem, etc.). These lines are frequently long and are often exposed to sources of electrical interference, such as relays, transformers, motors, solenoids, and even lightning. Electromagnetic interference (EMI) from these sources can cause the transmission of faulty data, overstressing the interface circuits, and (especially in the case of lightning) cause component failures. Generally, output line driver circuits tend to have higher-than-average failure rates, due both to EMI stressing and to the high transition currents that result from driving capacitative interfacing cables.

When I/O circuits are multiplexed (scanned in sequence), the interaction between common scan circuits must be considered in making a troubleshooting diagnosis. For example, in a terminal where the keyboard and display are scanned, a bad display driver can cause what appears to be a keyboard problem. Similarly, a stuck key can appear to make the display fail.

5-1.8 Verifying the Failure

As discussed in Chapter 1, always make sure that a problem really exists before starting any troubleshooting analysis. This means that you must understand the equipment. Microprocessors allow designers to design products that are not only complex in function, but sometimes complex in operation as well. Be sure the apparent problem is not a user error, but a real product malfunction. Few things are more frustrating than trying to fix something that is not broken. In some situations, it appears that a product should do something that it was not actually designed to do.

Design "bugs" in the firmware (usually the ROM) can sometimes cause failures when used under operating conditions that were not anticipated during product design. These are more likely to occur in early production runs and can best be verified (if suspected) by contacting the manufacturer. At the other extreme, a problem may actually exist but not show up because the product is not adequately exercised. These kinds of problems are often very simple to find (by observing a burned-out OHMs LED indicator when pushing the OHMs button on a DVM). Such problems can also be complex. For example, errors can occur when an unusual sequence of operations is performed. Because the complex problems are much more difficult to check, extensive test procedures are used to test products at the factory.

5-1.9 Milking the Front Panel

Many microprocessor-based products have some sort of front panel complete with controls and indicators. A great deal of troubleshooting data can be obtained from such products by operating the controls and observing the indicators. (This is known as "milking the front panel.") For example, if the indicators are all dead when the power is turned on, you might suspect a bad switch, fuse, power cord, battery connection, or power supply. If one segment of a digital display is dead, with other segments on, the problem is probably the display itself or the circuit that drives the display. If the only failure of a DVM is in the range 1 to 10 V, the problem area can be narrowed down to a relative small portion of the circuit (probably the attenuator).

5-1.10 Checking the Service Literature

Always take advantage of any designed-in performance verification or power-up test modes and diagnostic messages you find in the service literature. You may find many service aids and procedures in the manual just waiting for you to try. Special service switches, jumpers, test fixtures, indicators, and test techniques can make the job much easier. Try to understand the circuits and figure out where the major components are located. Check the theory of operation section first, comparing what you read against the block diagrams and schematics. Usually, you do not have to read every boring detail of the manual, but you must have a good idea of circuit operation. As a minimum for any microprocessor-based equipment, you must be able to identify the microprocessor, ROM, RAM, I/O, address decoder, clock, bus, control or enable, and interrupt portions of the system.

5-1.11 Production Versus Field Failures

Common faults and the best troubleshooting techniques for finding them depend on product history. When a new product is first turned on at the factory, almost anything can go wrong. Products that fail in the field have all worked at one time. Assembly errors, such as misloaded components and miswired circuits, generally need not be considered in field failures. Also, the likelihood of solder shorts and multiple faults is much greater on the production line than in the field. Field failures are usually caused by components or connections that have failed.

5-1.12 Some Common Production-Line Troubleshooting Problems

A number of production-line troubleshooting problems can be found by looking for the obvious. Start by looking for the easy things to test and repair. Simple things are as likely to fail as the complicated ones. For example, the power supply is one of the more failure-prone portions of many products, and is one of the easiest to test (and is usually simple to troubleshoot). An out-of-spec voltage can cause erratic circuit performance. If you do not check the voltages first, it could take considerable time to find the problem.

A thorough mechanical or visual inspection should always be performed early in the troubleshooting sequence. Poor PC board and cable connections, broken wires, and loose parts can usually be found either visually or by touch. Look for improperly set switches, improperly connected jumpers, misloaded components (wrong ones and backward ones), and cold solder joints. Backward resistor packs can be particularly hard to diagnose electrically because they can cause interaction between unrelated logic elements, but such packs are usually easy to check visually.

Solder and gold (copper) shorts on PC boards are common failures in production-line equipment. (Generally, such problems do not appear in the field unless there has been some repair.) Shorts can sometimes be removed with a sharp knife. When the precise location of a short is not known, there is a technique (not always recommended) for removing the short. The technique is sometimes useful for situations in which the location of the short is not accessible (such as inner layer shorts on multilayer PC boards). The procedure involves charging a $100,000\text{-}\mu F$ (or larger) capacitor to 5 V (a safe voltage for most logic circuits). Then, with cables solidly connected to the two shorted nodes and proper polarity observed, discharge the capacitor into them and listen for a snapping sound on the board. Check continuity to see if the short has been opened, and, if not, try again. This technique should be used with caution since it will open the weakest link of the current path, which may not always be the fault source, but may be a fine trace or a plate-through on the board. The current tracer described in Chapter 4 provides a much safer means for finding shorts, if the short is accessible to the tracer.

When automatic component insertion equipment is used on the production line, the problem of bent-under IC pins is common. Such bent-under pins can result in an open electrical connection between the IC and the PC board, and intermittent connection, or shorts to traces near or under the IC. The bent-under pin is often difficult to spot visually because it may look as though the pin is properly soldered in place. The best way to tell is to look at the bottom of the board for the ends of any IC pins, or along the plane of the board to see under the ICs.

Edge connectors are commonly used in PC boards. Such connectors may cause problems in production when their borders are cut off-center, or when they are accidentally covered with board sealing spray or solder resist. Such problems will usually show up during visual inspection. Multilayer PC boards suffer from all the problems of regular boards, plus some special problems of their own. Misregistration and contamination of inner layers (which can cause high-frequency or leakage problems) can often be observed by holding the board up to the light. Since repair of the inner layers is often impossible, the entire board may have to be replaced. Wire-wrap boards are prone to bent posts that cause shorting. Other common production problems include 14-pin ICs loaded into the wrong end of a 16-pin socket, miswiring, wire shorts between pins, and signal coupling (crosstalk) due to closely bundled wires.

5-1.13 Mechanical Failures in the Field

A visual inspection of equipment that fails in the field can often reveal such things as loose wires, broken traces, cracked ceramic ICs, and resistor packs, bent wire-wrap posts, and dirty connectors. A "calibrated fist" on the side of the cabinet can often be used to detect loose or intermittent connections and stuck relays. Mechanically stressing boards and connectors (by twisting and flexing) can often help to locate some of these problems.

When a product fails on the first try in the field, or has been subjected to rough handling, look for problems with the PC board edge connectors. Try reseating all the assemblies and circuit board connections to determine if the problem is one of poor connector contact. A pencil eraser is useful for cleaning dirty edge connectors.

5-1.14 Board Swapping

If any of the PC boards are easy to replace, and known good boards are at hand, you can try swapping the boards. When duplicates of the same board or assembly are used on one product, they can be swapped with each other. The risk involved in board swapping is that you could damage a good board because of the same electrical overload that damaged the bad one when it was installed. In any case, power to the equipment should be turned off when removing or installing boards or assemblies.

If an identical product is available, functional comparisons can sometimes be informative. This comparison can be especially useful in situations where it is not clear that there is actually a hardware problem (it may be an equipment idiosyncrasy or design limitation).

If a device in a socket is suspect, try tapping (not pounding) the device to see if there is a loose connection. Then try substituting a known good device. Note that one of the last devices you should suspect, but that is most often the first to be replaced, is the microprocessor. The actual failure rate for microprocessors is very low. However, because they are complex and their correct operation is difficult to verify, microprocessors are often the first to be removed from a PC board. This is also true of the LSI chips used with microprocessors.

5-1.15 Stress Testing

Stress testing can be very effective in dealing with marginal or intermittent failures, and can often cause these types of failures to temporarily improve or deteriorate. Either way, a stress test can sometimes locate a fault more quickly than by circuit test and analysis. Boards are stressed physically by tapping or twisting them, thermally by heat (air gun or hair dryer), or by cooling them (from an aerosol freeze can), and electrically by varying the supply voltage. Thermal stressing can be used to isolate a fault in a specific device on a board more precisely than the other methods because heat or cold can be applied directly to a single component. Intermittents can result from marginal chips, lead bonds, solder joints, connections, and drive or timing circuits.

Briefly touching each device on a circuit board can pinpoint a component that is running hot (much hotter than the others). When a particular device runs significantly hotter than others of the same type, a problem may exist. A faulty device can sometimes be hot enough to burn your finger, so use this technique with caution! Also, be aware that some good devices may run hotter than you expect during normal operations, and that temperatures may vary widely from one device to another.

5-1.16 Power-Supply Shorts

There are some effective ways of dealing with shorts across the power supply. The first thing to do in a multiboard system is to try to localize the short to a single board. This can be done by removing one board at a time until the power supply is no longer shorted. The last board to be removed is the shorted one.

One technique for finding the short on a faulty board is to inject current through the two shorted lines with the logic pulser. The current tracer is then used to follow this current to the short. Similar procedures are described in Sec. 5-2. Keep in mind that capacitors (especially electrolytics) have some current going into them because of the pulsing current. Shorted capacitors can be found by using the current tracer to compare the current levels going into identical capacitors on the same board. The capacitor that shows a much higher level than the others is likely to be shorted. This technique is particularly useful for finding shorted ceramic bypass capacitors.

Another technique for locating power bus shorts is to supply a relatively high current (about 3 to 5 A) into the short. Be sure to maintain the same voltage polarity and not to exceed the supply voltage normally present. The current path to the short can often be determined by using a DVM with high resolution (0.01 μV or better) to look at voltage drops on the power bus traces. Voltages are developed across the traces in the path going to the short, but not in other paths, as shown in Fig. 5-2.

Another method often used in the days before current tracers were available is to freeze the entire board (to about $-10°C$), allow moisture to condense on the board, and then power the board with a 3- to 5-A supply (but at the normal supply voltage). As the board warms up and defrosts, the current path becomes visible and, in many cases, will pinpoint the short. Although this

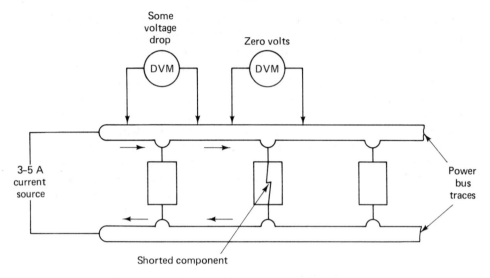

Figure 5-2 Using a sensitive voltmeter to locate power bus shorts.

technique is recommended by Hewlett-Packard, and author recommends that
the method be used only where the power bus is inaccessible.

5-1.17 Isolating Faults by Analyzing Test Results

Once the easy things have been tried unsuccessfully, it is necessary to start
analyzing test results. At this point, individual troubleshooting skills, intuition,
and knowledge of the equipment really make a difference.

First, be sure to take advantage of any designed-in and documented circuit isolation features, such as selected board removal, service jumpers, and
special test modes and procedures. It can be very useful to separate the
microprocessor system from the peripheral circuits to allow diagnosis of each
portion separately.

The half-split technique described in Sec. 1-5.5 can be used effectively at
this point, if the microprocessor system is both digital and analog, as is often
the case. As discussed, half-splitting involves choosing a point roughly in the
middle of the circuit, and checking if the fault exists before or after the selected
point. This process works best in circuits that have clear, unidirectional signal
paths without large feedback loops. Even with microprocessor-based systems,
this approach can be effective because the circuits outside the microprocessor
portion often fit these guidelines.

In typical equipment, the first half-split is generally done at the digital-to-
analog interface, if possible. Analog circuits often have higher failure rates (due
to higher demands made on speed, power, temperature, sensitivity, accuracy,
adjustment, external overloads, and reduced component safety margins). Also,
the analog circuits often outnumber the digital circuits. There is also a possibil-

ity of electrical interaction of clock and TTL power bus lines with analog circuits, which can cause serious system noise problems.

One way to make a quick half-split to isolate the digital portion of any microprocessor-based system is to check for signal activity. Using a logic probe, you can examine activity on the clock, reset, control, and chip enables lines, as well as the data and access buses. If you do not find any activity on these lines, with the equipment supposedly running, you have definitely isolated problems to the digital portion of the system.

5-1.18 Digital Failure Modes

The most common failure mode for digital ICs is open lead bonds inside the package. This problem is discussed further in Sec. 5-2. There are thin wires connecting the package pins to the IC chip. If an output lead bond opens, the output pin floats and the logic probe probably indicates a constant floating logic level because of other device inputs connected to that node. If an input lead bond opens, one or more outputs of that IC usually appear to malfunction (stuck high, low, or executing the logic function incorrectly). If any of these outputs goes to a three-state bus, it can cause bus conflicts (more than one output on at a given time, also known as a "bus fight"). As discussed in Sec. 5-2, the current tracer is a most effective tool for finding the causes of a bus fight.

Another common digital IC failure is a shorted input pin to ground. This fault is often caused by a bad input protection diode on the chip, and appears as a stuck low level. An oscilloscope connected to a node with this type of problem shows a voltage level near ground being pulled up, perhaps a few hundred millivolts, whenever a logic 1 output on that node turns on. Again, the current tracer is the best instrument for finding such problems.

If a current tracer is not available, another means for locating stuck inputs and outputs involves the use of a sensitive, high-resolution DVM and a can of cold spray. Connect the DVM to the stuck node and select the most sensitive d-c voltage range available. Then, while monitoring the voltage, spray each IC connected to the stuck node, one at a time, to change the IC temperature. Any noticeable change in voltage (more than 10 mV) on the node indicates that the IC being sprayed is drawing current. If a freeze can is not available, a heat source can be used instead. This technique relies on the properties of the semiconductor material used in the IC that relates voltage (or current) to temperature. Typically, when silicon is increased in temperature, current passing through junctions in the silicon also increases.

5-1.19 Some Practical Isolation Techniques

Once a particular input or output pin is suspected, it is useful to isolate the pin from the rest of the circuit. A quick, nondestructive way to do so is to suck the solder away from the area between the pin and the PC board pad, using a vacuum desoldering tool (solder gobbler) or solder wicking braid. Then bend the pin so that it is centered in the pad hole, not touching the pad at any point. Use a continuity tester or ohmmeter to verify that the pin is no longer in electrical contact with the board.

The techniques that you can use to isolate the digital blocks of microprocessor-based equipment are entirely dependent on the electrical and mechanical architecture of the system. For example, if some of the digital boards can be removed and still allow the basic microprocessor and RAM/ROM (or system *kernel*) to operate, the boards should be removed to see if the fault is cleared. If the kernel can be allowed to run open-loop (no feedback from the data bus) a free-run mode can sometimes be used to check the kernal and address bus activity. This subject is discussed further in Sec. 5-3.

An extender board with switches on bus and signal lines can be used to break selected signals between a PC board and the rest of the system. Thus, feedback paths and stuck buses can be removed from the main system.

An even simpler way to open selected signals going through a board edge connector is to place a tape or stiff paper on the PC board edge fingers that you wish to isolate. Be sure to make a record so that you will remember to remove the tape or paper when the test is completed.

An often effective means of detecting bus line problems is to measure the resistance to ground (with the power off) of each line in the bus (data bus, address bus, etc.). The resistance of each of these lines is usually the same. If any one differs substantially, you may suspect a problem on this line. If two lines show the same (higher or lower) resistance, the two lines may be shorted together. In either case, before going further, check the schematic to see if the arrangement of circuits connected to these lines could explain the differences.

Overriding interrupt lines and chip enable pins on suspected devices can be used to verify that the IC is functioning correctly. This can be done by momentarily shorting the appropriate pin high or low or, preferably, by using the logic pulser as described in Sec. 4-5.

5-1.20 Feedback Loop Problems

Digital feedback loops are often difficult to troubleshoot because errors propagate around and around. A feedback loop with a faulty output signal sends this signal back to the input to produce more bad outputs. Opening this feedback path prevents the fault output signals from going back to the input. Then, if controlled inputs to the loop can be generated, the signal flow from the input to the output can be observed. However, it is not always easy to provide this input (many lines may need to be controlled). It may also be difficult to predict correct circuit operation. The feedback problem is discussed in great detail in Sec. 5-3.

If other working equipment (or a board with the same circuitry) is available, it is sometimes practical to allow the output of the good circuit to control the inputs of both circuits (suspected and known good circuits). In this manner, you know that the circuit under test is getting the correct input signal. It is then a matter of comparing the nodes of the two circuits and looking for differences. A signature analyzer can be useful for making this comparison.

5-1.21 Comparison of ICs

A technique that can sometimes be used to locate defective ICs is called "piggy-backing." The piggy-backing technique involves looking at suspected IC outputs with an oscilloscope or signature analyzer and then placing an identical IC package directly on top of the IC under test. The pins of the known-good or piggy-back IC must usually be bent slightly to ensure that all pins of both ICs are in good electrical contact. If no change is observed and the outputs are not stuck, it can generally be assumed that the IC in question is not the problem. Be cautious of ICs with sequential circuits (such as shift registers and counters) that may cause output differences because of startup conditions. A better way of performing the piggy-back test is to use an IC comparator as described in Sec. 4-8. The comparator provides an instant indication of differences between the two ICs.

5-2. BASIC DIGITAL IC TEST AND TROUBLESHOOTING

The circuits of a typical microprocessor-based system are mostly IC rather than discrete. As in the case of a linear or analog IC, you do not have access to the internal circuits of a digital IC. Thus, you must determine if the IC is performing its function, or failing, without actually getting into the circuits. (From a practical standpoint, it is very helpful if you can make this decision before you pull the IC from the PC board!) Unfortunately, the traditional tools for discrete digital troubleshooting often provide little or no help in making these decisions in the case of digital ICs. Let us consider two classic examples of digital troubleshooting techniques, the comparison of timing functions, and point-to-point signal pulse tracing.

5-2.1 Timing in Digital IC Troubleshooting

An often-used method for troubleshooting discrete digital circuits is to introduce a pulse train at the circuit input while monitoring various points throughout the circuit with an oscilloscope. If any of the pulses are absent or abnormal (such as very low amplitude pulses or pulses that occur at the wrong time), this means a failure or degrading of the components between the pulse source (pulse generator) and monitor (oscilloscope). For example, a degraded diode or transistor can introduce an abnormally long delay and cause an output pulse to occur too late to open a buffer. A leaking capacitor can appear as "pulse jitter" on the oscilloscope. However, when a digital IC fails, there is usually total failure. Digital IC timing parameters rarely degrade or become marginal after prolonged use. Thus, critical timing measurements are usually not too important in troubleshooting.

This does not mean that timing measurements are to be omitted from digital troubleshooting. Timing measurements are often critical during the design and development stage of digital systems. Generally, it is a case where some IC (or group of ICs) does not operate fast enough to perform each program step, as initially programmed. Such problems are solved by debugging (changing the program sequence) rather than by troubleshooting. As a general rule, once a program has been debugged and the digital device performs properly, timing will remain good (or will fail completely).

There are exceptions to this general rule. Digital ICs tend to speed up when power-supply voltages are high, and slow down with low power-supply voltages. Thus, if program timing is on the borderline, and there is a drastic change in power-supply voltage (particularly if the voltage is low), improper timing can result. Digital ICs are also affected by temperature extremes. Such temperature extremes could possibly affect digital systems with marginal timing.

5-2.2 Point-to-Point Pulse Tracing

Basic point-to-point pulse tracing can be used in digital troubleshooting. Pulses are introduced on a line at some point in the circuit (with a pulse generator or logic pulser), and then monitored at another point on the line (with an oscilloscope, logic probe, or clip). Unfortunately, there are certain cases where this does not work too well with digital ICs.

Consider the circuit of Fig. 5-3, which is part of the output buffer (connected to one line of the data bus) in a microcomputer RAM. (The circuit is sometimes referred to as a "transistor totem pole.") In either the high or low (1 or 0) state, the circuit is a low impedance. In the low state, the circuit output is a saturated transistor to ground, and appears as about 5 or 10 Ω to ground. This presents a problem in pulse tracing. A signal source used to inject a pulse at a point that is driven by this output (such as trying to inject a pulse on that line of data bus) must have sufficient power to override the low-impedance output state.

A typical pulse generator (even a logic pulser) may not have this capability. Instead, it is necessary to either cut PC traces (or lines) or pull IC leads in order to inject the pulses. Both of these practices are time consuming and lead to unreliable repairs. However, there are basic digital IC troubleshooting techniques, using logic probes and pulsers, as well as current tracers, that will resolve such problems. These techniques are discussed in the following paragraphs. But first let us consider the how and why of digital IC failure.

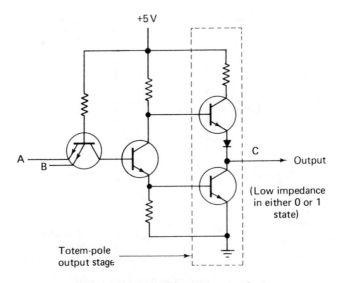

+5 V

A
B
C
Output

(Low impedance
in either 0 or 1
state)

Totem-pole
output stage

Figure 5-3   Output buffer with totem-pole output.

5-2.3 How Digital ICs Fail

It is essential that you understand the types of failure that occur in digital ICs if you are going to troubleshoot a digital device effectively. There are two classes of digital IC failure: (1) those caused by internal failure, and (2) those caused by a failure in the digital circuit external to the IC.

Four types of failure can occur internally to a digital IC: (1) an open bond on either the input or output, (2) a short between an input or output and V_{CC} or ground, (3) a short between two pins (neither of which is V_{CC} or ground, and (4) a failure in the internal circuitry (often called the steering circuitry) of the IC.

Four types of failure can occur in the circuit external to the IC: (1) a short between a point or node and V_{CC} or ground, (2) a short between two nodes (neither of which is V_{CC} or ground), (3) an open signal path, and (4) a failure of an external component.

Effects of an open output bond.   When there is an open output bond, as shown in Fig. 5-4, the inputs driven by that output are left to float. In a typical TTL IC, a floating input rises to approximately 1.4 to 1.5 V, and usually has the same effect on circuit operation as a high logic level. Thus, an open output bond will cause all inputs driven by that output to float to a bad level, since 1.5 V is less than the typical TTL high threshold level of 2 V, and greater than the low threshold level of 0.4 V. In TTL ICs, a floating input is usually interpreted as being a high level. The effect is that these inputs respond to the bad level as though it were a static high level.

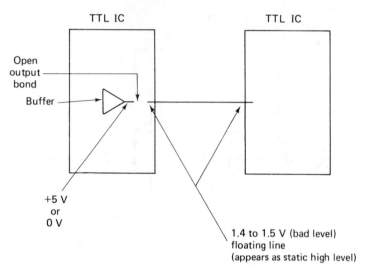

Figure 5-4   Effects of an open output bond.

Effects of an open input bond.   When there is an open input bond, as shown in Fig. 5-5, the open circuit blocks the signal driving the input from entering the circuit (a microprocessor data bus buffer, in this case). The input line is thus allowed to float, and responds as though the input is a static high signal. It is important to realize that, since the open occurs on the input side of the IC, the digital signal driving this input is unaffected by the open. That is, the signal is blocked to the microprocessor buffer inside the IC, but there is no effect on other ICs connected to the same line. Of course, the microprocessor responds as if there is a static 1 on that line of the data bus, and thus does not

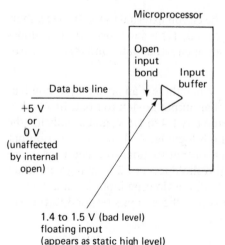

Figure 5-5   Effects of an open input bond.

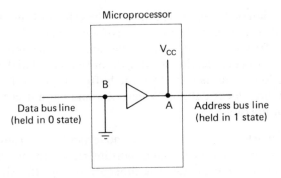

Figure 5-6   Effects of a short between an input or an output and V_{CC} or ground.

respond properly to those commands or words where a 0 is to appear on that line of the bus.

Short between an input or output and V_{cc} or ground.   A short, as shown in Fig. 5–6, has the effect of holding all signal lines connected to that input or output either high (in the case of a short to V_{cc}) or low (if shorted to ground). In the case of Fig. 5–6, the address line connected to point A is held in the high (1) state, and the data line connected to point B is held low (0). This results in a total disruption of the program and is thus one of the easiest types of digital IC failure to locate.

Short between two pins.   Shorts such as that shown in Fig. 5–7 are not as straightforward to analyze as the short to V_{cc} or ground. When two pins are shorted, the outputs driving those pins (the ROM and RAM in this case) oppose each other when one attempts to pull the pins high while the other attempts to pull them low. In this situation, the output attempting to go high supplies current, while the output attempting to go low sinks this current. Whenever both outputs attempt to go high simultaneously or to go low simultaneously, the shorted pins respond properly. But whenever one output attempts to go low, the short is held low.

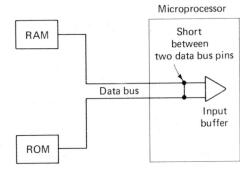

Figure 5-7   Effect of a short between two pins.

Failure of internal control circuitry. The effects of internal failure are difficult to predict. For example, in the case of Fig. 5–8, a failure of the internal circuit has the effect of permanently turning on either the upper transistor of the buffer output totem pole, thus locking the buffer output in the high state, or turning on the lower transistor to lock the buffer output in the low state. Of course, this failure blocks data bus signal flow and has a disastrous effect on microprocessor operation.

External shorts. A short between a node and V_{CC} or ground, external to the IC, cannot be distinguished from a short internal to the IC. Both cause the signal lines connected to the node to be either always high (for shorts to V_{CC}) or always low (for shorts to ground). When this type of failure is found, only a very close physical examination of the circuit can show if the failure is external to the IC. The current tracer is an effective tool for locating such shorts.

Open signal paths in circuits. An open signal path in the external circuit has an effect similar to that of an open output bond driving the node. As shown in Fig. 5–9, the input (microprocessor) to the right of the open is allowed to float to a bad level and thus appears as a static high level in typical TTL operation. Those inputs to the left of the open (data bus) are unaffected by the open and thus respond as expected.

5-2.4 Examples of Digital Troubleshooting

Practical digital troubleshooting is a combination of detective work or logical thinking and step-by-step measurements. Thus far in this chapter and in Chapter 4, we have described the test equipment that is available for digital

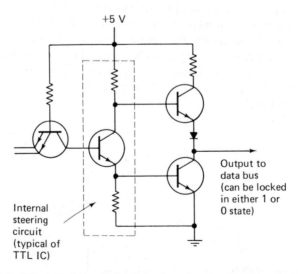

Figure 5–8   Effect of a failure in the internal steering circuitry of a typical TTL IC.

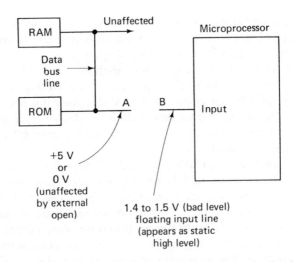

Figure 5-9   Effect of an open signal path in external circuits (data bus).

work and how to use the equipment effectively in testing digital circuits. The following paragraphs of this section describe the final step: combining all of these practical techniques with logical thinking to solve some problems in digital troubleshooting.

Two typical digital circuits are discussed, one very simple (the half-adder) and one more complex (a decade counter and readout). All three aspects of digital troubleshooting are included: self-check, pulse measurement, and logical thinking. Before going into the specific examples, let us review some very practical details of digital troubleshooting.

5-2.5 Practical Digital Troubleshooting Techniques

The first step in any troubleshooting process is to narrow the malfunctioning area as much as possible by examining the observable characteristics of failure. For example, in the case of a microcomputer, this often involves punching in data at the keyboard and noting the response on the CRT screen. Furthermore, the designers of some microcomputers and other programmed devices have developed diagnostic routines or short programs that isolate troubles down to specific ICs or groups of ICs. (Section 5-3 describes some examples of simple diagnostic routines for a microprocessor-based system.) The routines found in the service literature should be followed religiously.

In the absence of good diagnostic routines, and when CRT responses are meaningless, the logic analyzer described in Chapter 4 is the most effective tool for observing characteristics of failure in programmed digital equipment. As discussed, a logic analyzer makes it possible to observe the full transfer of data in a programmed system (such as a microcomputer) for each step of the program. That is, you can see and compare the data word (on the data bus) at each address (on the address bus) for all program steps. This can be done one step at a time, in groups of steps, or you can move quickly through the program steps to an area of the program where trouble is suspected.

Based on observations of the logic analyzer (or CRT/keyboard and diagnostic routines), you can localize the failure to as few ICs or other circuits as possible (hopefully to one circuit or IC). At this point it is necessary to narrow the failure further to one suspected circuit by looking for improper key signals or pulses between circuits. The logic probe is a most effective tool for tracing key signals in digital circuits.

In many cases, a signal will completely disappear (no clock signal, no read/write signals, etc.). By rapidly probing the interconnecting signal paths (clock line, read/write line, etc.), a missing signal can be readily found. Another important feature is the occurrence of a signal on a line that should not have a signal. The pulse memory option found on some probes allows such signal lines to be monitored for single-shot pulses or pulse activity over extended periods of time. The occurrence of a signal is stored and indicated on the pulse memory LED.

Isolating a failure to a single circuit requires knowledge of the digital equipment circuits and operating characteristics. In this regard, a well-written service manual is invaluable. Properly prepared manuals show where key signals are to be observed. The logic probe then provides a rapid means of observing the presence of these signals.

Once a failure has been isolated to a single circuit or IC, the logic probe, logic pulsers, and current tracer can be used to observe the effects of the failure on circuit operation and to localize the failure to its cause (either an IC or a fault in the circuit external to the IC). The logic clip and comparator can also be used, if they can accommodate the number of IC pins.

The classic approach at this point is to test the suspected IC (or ICs) using a logic comparator or by substitution. However, the following steps are based on the assumption that neither of these approaches is practical (the comparator will not accommodate the ICs, and the ICs are very difficult to remove).

Checking for pulse activity on a line.   The logic probe can be used to observe signal activity on inputs and to view the resulting output signals. From this information, a decision can be made as to the proper operation of the IC. For example, if a clock signal is occurring on the clock line to a RAM or ROM, and the enabling inputs (such as read or write signals, chip enable signals, etc.) are in the enabled state, there should be signal activity on the data bus lines. Each line should be shifting between high and low (1 or 0) as the program goes through each step. The logic probe allows the clock and enabling input to be observed and, if pulse activity is indicated on the outputs (each of the data bus lines, in this case), the ROM and RAM can be considered as operating. As discussed, it is usually not necessary to measure actual timing of the signals since ICs generally fail catastrophically. A possible exception is when an output buffer does not open and close at the proper time. However, with few exceptions, the occurrence of pulse activity is usually sufficient indication of operation. Of course, the ROM and RAM could have incorrect data bytes stored at various addresses, or the data bytes could be read out incorrectly to the data bus (due to open bonds, shorts, etc.).

When more detailed study is desired or when input signal activity is missing, the logic pulser can be used to inject input signals and the probe used to monitor the response. The logic pulser is also valuable for replacing the clock, thus allowing the circuit to be single-stepped, while the logic probe is used to observe changes in the output state (such as changes on the address and data bus lines).

When using this first step to compare inputs and outputs of the ROM, RAM, and so on, it is wise to check all the output lines before you try to check the first fault you find. Prematurely studying a single fault can result in overlooking faults that cause multiple failures, such as shorts between two lines in the data or address bus. This often leads to the needless replacement of a good IC and much wasted time. The extra work can be minimized by systematically eliminating the possible failures of digital circuits, as discussed.

Testing for an open bond.   The first failure to test for is an open bond in the IC driving the failed node. The logic probe provides a quick and accurate test for this failure. If the output is open, the node floats to a bad level. By probing the node, the logic probe quickly indicates a bad level. If a bad level is indicated, the IC driving the node is suspect.

Testing for shorts to V_{CC} or ground.   If the node is not at a bad level, test for a short to V_{CC} or ground. This can be done using the logic pulser and probe. (The current tracer can also be used effectively to find shorts, as discussed in Sec. 5-2.6.) Although the logic pulser is powerful enough to override even a low-impedance output, it is not powerful enough to effect a change in state on a V_{CC} or ground bus. Thus, if the logic pulser is used to inject a pulse while the logic probe is used simultaneously on the same node to observe the pulse, a short to V_{CC} or ground can be detected. The occurrence of a pulse indicates that the node is not shorted, and the absence of a pulse indicates that the node is shorted to V_{CC} (if the indication remains high) or ground (if the indication remains low).

Cause of V_{CC} or ground shorts.   If the node is shorted to V_{CC} or ground, there are two possible causes. The first is a short in the circuit external to the ICs, and the other is a short internal to one of the ICs attached to the node (as discussed in Sec. 5-2.3). The external short should be detected by an examination of the circuit. If no external short is found, the cause is likely to be any one of the ICs attached to the node. These can be eliminated one at a time. Also check for shorted capacitors and resistors on the line. (Note that resistors do not usually short, but their terminals can be shorted by mechanical vibration.)

Testing for shorts between nodes.   If the node is not shorted to V_{CC} or ground, or is not an open output bond, look for a short between two nodes. This can be done in one of two ways. First, the logic pulser can be used to pulse the failing node being studied, and the logic probe can be used to observe each of the remaining failing nodes. If a short exists between the node being studied and one of the other failing nodes, the pulser causes the node being probed to change state (the probe will detect a pulse). To ensure that a short exists, the probe and pulser should be reversed and this test made again. If a pulse is again detected, a short is definitely indicated.

Causes of shorts between nodes.   If the failure is a short, there are two possible causes. The more likely cause is a problem in the circuit external to the IC. This can be detected by physically examining the circuit and repairing any solder bridges or loose wire shorts found. Only if the two shorted nodes are common to one IC can the failure be internal to that IC. If after examining the circuit, no short can be found external to the IC, the IC should be replaced.

Testing for an open signal path.   If no nodes are indicated as failing, but there is a definite failure of the circuit, an open signal path can be suspected. From a practical standpoint, it is generally easier to check for open signal paths than to replace ICs. The logic probe provides a rapid means of not only detecting but also physically locating an open in a circuit, such as that shown in Fig. 5-10, since an open signal path allows the input to the "right" of the open (point B) to float to a bad level. Once an input floating at a bad level is detected, the logic probe can be used to follow the circuit back from the input, while

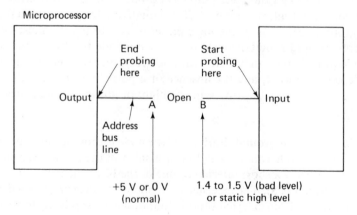

Figure 5-10   Tracing an open signal path in an address bus line.

looking for the open. This can be done because the circuit to the "left" of the open (point A) is at a good logic level (either high, low, or pulsing), while the circuit to the right (point B) is at a bad level. Probing back along the signal path (from right to left) indicates a bad level until the open is passed. Thus, the probe finds the precise physical location of the open.

5-2.6 Digital Troubleshooting with the Current Tracer

The basic procedures for using the current tracer are discussed in Sec. 4-6. The following paragraphs describe some additional uses for the current tracer in digital troubleshooting. From a practical standpoint, whenever a low-impedance fault exists, whether on a digital board or not, the shorted node can be stimulated with a logic pulser and the current followed by means of the current tracer.

Ground planes.   Occasionally, defective ground planes can cause problems in digital circuits. It is possible to determine the effectiveness of a ground plane by tracing current distribution through the plane. This is done by injecting pulse current into the plane from a logic pulser (or any pulse generator) and tracing the current flow over the plane. Generally, current flow should be even over the entire plane. However, it is possible that the current will flow only in a few paths, particularly along the edges of the ground plane.

V_{CC}-to-ground shorts.   The existence of a V_{CC}-to-ground short is generally easy to determine. Typically, there is a drop in power-supply voltage or a complete failure of the power supply. However, locating the exact point of the short can be quite difficult. To use the current tracer in this application, disconnect the power supply from the V_{CC} line, and pulse the power-supply terminal using the logic pulser with the supply return connected to the ground lead of the pulser. Even if capacitors are connected between V_{CC} and ground, the current tracer usually shows the path carrying the greatest current.

Stuck line caused by dead driver.   Figure 5-11 shows a frequently occurring trouble symptom: a line (say in the address or data bus) has been identified on which the signal is stuck high or low. The particular line is suspect since it shows no pulse activity (as discussed in Sec. 5-2.5). Is the driver (or the signal source) dead, or is something, such as a short, clamping the line to a fixed value (such as to ground or V_{CC})? This question is readily answered by tracing current from the driver to other points on that line. If the driver (source) is dead, the only current indicated by the tracer is that caused by parasitic coupling from any nearby currents, and this is much smaller than the normal current capability of the driver. On the other hand, if the driver is good, normal short-circuit current is present and can be traced to the short or element clamping the line.

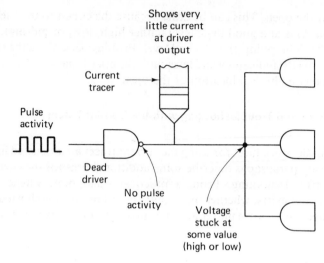

Figure 5-11　Tracing an open signal path in an address or data bus line.

Stuck line caused by input short.　　Figure 5–12 illustrates this situation, which has exactly the same voltage symptoms as the previous case of a stuck line caused by a dead driver. However, the current tracer now indicates a large current flowing from the driver, and also makes it possible to follow this current to the cause of the problem (a shorted input in this case). The same procedure also

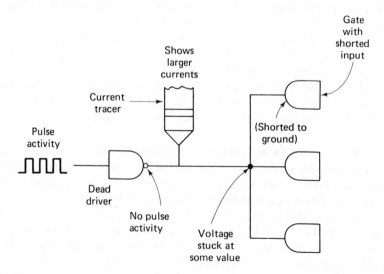

Figure 5-12　Using current tracer to show that stuck node is caused by an input short.

finds the fault when the short is on the interconnecting line (for example, a solder bridge to another line).

Stuck three-state data or address bus.   A stuck three-state bus, such as a microprocessor data or address bus, can present a very difficult trouble-shooting problem, especially to voltage-sensing measurement tools (such as a logic probe). Because of the many bus terminals (typically 8 or 16), and the fact that several ROMs and RAMs may be connected, it is very difficult to isolate the one element (ROM or RAM) holding the bus in a stuck condition. However, if the current tracer indicates high current at several outputs of a ROM or RAM, it is likely that one (and most likely only one) element is stuck in a low-impedance state. The defective element is located by placing the ROM or RAM control input line to the appropriate level for a high-impedance output state (the "off" condition), and noting whether high current flow persists at the ROM/RAM buffer output. This is repeated for each ROM/RAM until the bad one is located.

If the current tracer indicates high currents at only two elements (say two RAMs), the problem is a bus fight. That is, both RAMs are trying to drive the bus at the same time. This is probably caused by improper timing of control signals to the RAMs. (One RAM buffer is opened before the other RAM buffer is closed.)

If the current tracer indicates the absence of abnormally high current activity on all elements, yet the bus signals are known to be incorrect, the problem is an element stuck in the high-impedance state. This can be found by placing a low impedance on the bus, such as a short to ground, and using the current tracer to check for the element that fails to show high-current activity.

Current tracer problems.   Efficient use of the current tracer usually requires a longer familiarization period than does the operation of voltage-sensing instruments (such as the probe). This is primarily because most electronic technicians are used to thinking in terms of current and the information that current provides, simply because this information has not been available conveniently. Most troubleshooting techniques involve measurement of voltage and resistance. Without a current tracer, it is necessary to open a line and insert an ammeter. This is not practical for most situations, so current is determined by calculation (based on observed voltage and current).

Use of the current tracer also requires some skill to avoid the crosstalk problem. That is, if a small current is being traced in a conductor that is very close to another conductor carrying a large current, the sensor at the tip of the tracer may respond to the current in the nearby line. The sensor of the current tracer has been designed to minimize this effect, but crosstalk can never be entirely eliminated. However, by observing the variation of the current tracer's display as the tracer is moved along the line, the operator can learn to recognize interference or crosstalk from a nearby line.

5-2.7 Half-Adder Troubleshooting

Figure 5-13 shows the diagram and truth table for a half-adder circuit. Assume that this circuit is found in an older computer and is made up of replaceable gates mounted on a plug-in printed circuit card (none of which is typical of present-day equipment but was quite common a few years ago). The entire half-adder circuit can be replaced as a unit by replacing the plug-in card (as a field-service measure to get the computer operating immediately). Then the card can be repaired by replacing the defective gates) as a factory or shop repair procedure).

Further assume that the computer failed to solve a given mathematical equation during self-check, the trouble is localized to the half-adder card, the card is replaced, and the computer then performs its functional normally. This definitely isolates the problem to the half-adder card. Now assume that the card can be connected to a power source (to energize the gates) and that pulses of appropriate amplitude and polarity can be applied to the two inputs (addend or digit A, and augend or digit B). The sum and carry outputs as well as any other points in the circuit can be monitored with an oscilloscope or logic probe.

Some service shops have special test fixtures that mate with printed-circuit cards. This provides a mount for the cards and ready access to the terminals. In other cases, cards are serviced in the equipment by means of an extender. The card is removed from its socket, the extender is installed in the empty socket, and the card is installed on the extender. This maintains normal circuit operation but permits access to the card terminals and components on the cards. These arrangements for PC cards are shown in Fig. 5-14.

To test the half-adder of Fig. 5-13, inject a pulse (true) at the addend in-

Addend (Digit A)	Augend (Digit B)	Sum	Carry
0	0	0	0
0	1	1	0
1	0	1	0
1	1	0	1

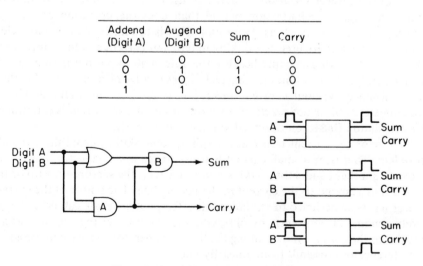

Figure 5-13   Half-adder troubleshooting.

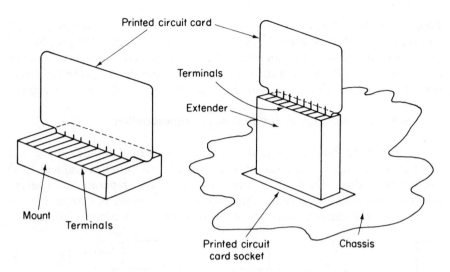

Figure 5-14   Printed-circuit card mount and extender.

put (digit A) and check for a true (pulse) condition at the sum output, as well as a false (no pulse) condition at the carry output. If the response is proper, both the OR gate and the B AND gate are functioning normally. To confirm this, inject a true (pulse) at the augend input (digit B) and check for a true at the sum output, as well as a false condition at the carry output.

If the response is not proper, either the OR gate or the B AND gate are the logical suspects. The A AND gate is probably not at fault, since it requires two inputs (digits A and B) to produce an output. To localize the problem further, inject a pulse at either addend or augend inputs and check for an output from the OR gate. If there is no output or the output is abnormal, the problem is in the OR gate. If the output is normal, the problem is probably in the B AND gate.

To complete the test of the half-adder circuit, inject simultaneous pulses at the addend and augend inputs, and check for a true (pulse) at the carry output and a false (no pulse) at the sum output. If the response is proper, all the gates can be considered to be functioning normally. If the response is not correct, the nature of the response can be analyzed to localize the fault.

For example, if the carry output is false (no matter what the condition of the sum output), the A AND gate produces a true output to the carry line when there are two true inputs. Since the test is made by injecting two true inputs, a false condition on the carry line points to a defective A AND gate. A possible exception is where there is a short in the carry line (possibly in the printed wiring). This can be checked by means of the logic probe, pulser, or current tracer, as described.

As another example, if the carry output is normal (true) but the sum output is also true, the B AND gate is the most likely suspect. The A AND gate produces a true output when both inputs are true. The B AND gate requires one true input from the OR gate and the false input from the A AND gate to produce a true condition at the sum output (because of the inversion at the input of the B AND gate). If the input to the B AND gate from the A AND gate is true, but the sum output shows a true condition, a defective B AND gate is indicated.

5-2.8 Decade Counter and Readout Troubleshooting

Figure 5–15 shows the logic diagram for a three-digit decade counter and readout. Each digit is displayed by means of a separate seven-segment LED display or readout. Each LED display is driven by a separate BCD seven-segment decoder and storage IC. The three decoder/storage ICs are enabled by a clock pulse at regular intervals or on demand. Each decoder/storage IC

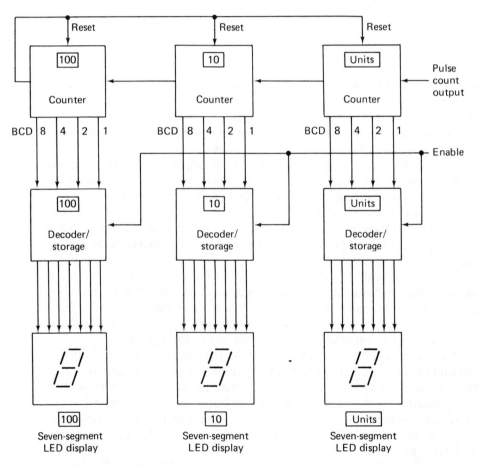

Figure 5-15 Troubleshooting diagram for three-digit decade counter and LED readout.

receives BCD information from a separate decade counter IC. Each counter IC contains four FFs and produces a BCD output that corresponds to the number of input pulses occurring between reset pulses. The maximum readout possible is 999. At a 1000 count, the output pulse of the 100 decade IC is applied to all three counter ICs simultaneously as a reset pulse.

Troubleshooting problem 1.   For the first problem, assume that the pulse input is applied through a gate and that the gate is held open for 1 s. The count shown on the LED readouts then indicates the frequency. For example, if the count is 388, this shows that there are 388 pulses passing in 1 s, and the frequency is 388 Hz.

Assume that a 700-Hz pulse train is applied to the input. The LED display should go from 000 to 700. However, assume that the displayed count is 000, 001, 002, 003, 004, 005, 000, 001, 002, 003, and so on. That is, the displayed count never reaches 006.

One logical conclusion here is that the problem is in the counter function rather than in the readout function (decoder / storage or LED). For example, if the units LED readout is defective, so that there is no display beyond 005, the 10 and 100 LED readouts are unaffected, producing a count of 000, 001, 002, 003, 004, 005, 010, 011, and so on.

With the problem localized to the counter function, there are three logical possibilities: the input pulses never reach more than five (not likely); there is a reset pulse occurring at the same time as the sixth input pulse or any time after the fifth input pulse (possible); and the units counter simply does not count beyond five (most likely).

With the detective work out of the way, the next practical step is to make measurements. To confirm or deny the reset pulse possibility, monitor the input pulse line and the reset line of the units counter on a dual-trace oscilloscope as shown in Fig. 5-16. Adjust the oscilloscope sweep frequency so that about 10 input pulses are displayed.

If a reset pulse does occur after the fifth pulse, the problem is pinpointed. Trace the reset line to the source of the unwanted (improperly timed) reset pulse.

If there is no reset pulse before the sixth pulse, monitor the input line and the 4-line of the units counter as shown in Fig. 5-17. The 4-line should go true at the sixth input pulse (as should the 2-line). Then monitor the 8-line. In all probability, the units counter is defective: the 4- and 2-lines show no output to the units decoder / storage IC, or the output is abnormal (pulses from the counter are below the threshold of the decoder / storage IC).

If a logic probe or logic clip is available, check operation of the units decade counter IC on a single-stepping basis. That is, inject input pulses and check that the 4- and 2-lines go true when the sixth input pulse is injected, as shown in Fig. 5-18. If circuit conditions make it possible, disconnect the 4- and 2-lines and recheck operation of the units counter. It is possible that the 4- or

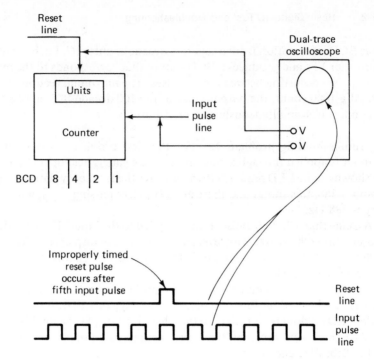

Figure 5–16 Checking if improperly timed reset pulse occurs after the fifth input pulse.

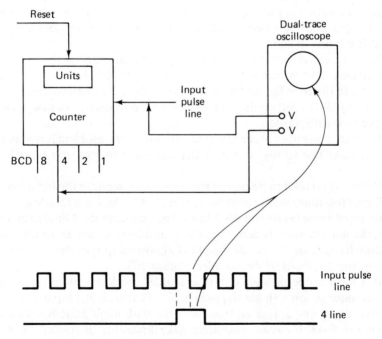

Figure 5–17 Checking if 4-line is true at the sixth input pulse.

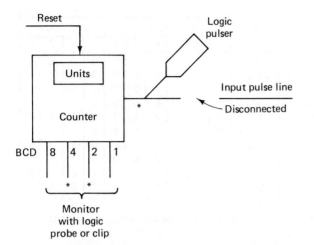

Figure 5-18   Checking units decade counter on a single-stepping basis.

2-line is shorted or otherwise defective. Unfortunately, with most present-day digital circuits, the lines or wiring to the ICs are in the form of printed circuits, making it impractical to disconnect individual lines or leads. The IC package must be checked by substitution (or by comparison if a logic comparator is available for the IC).

Troubleshooting problem 2.   Assume that the test conditions are the same as for problem 1. However, the readout is now 000, 001, 002, 003, 004, 005, 006, 007, 000, 001, 010, 011, 012, and so on. That is, the 008 and 009 displays are not correct.

The logical conclusion here is that the problem is in the readout function rather than in the counter function. For example, if there is a failure in the counter, the 10 counter never receives an input from the units counter. Since there is some readout from the 10 counter, it can be assumed that the units counter is functioning.

Assuming that the readout is faulty, there are several logical possibilities: the units counter output lines can be shorted or broken (the units decoder receives no input or an abnormal input); the units decoder output lines can be shorted or broken (thus the LED receives an abnormal input); or the units LED is defective.

The first practical step depends on the test equipment available. Monitor the units decoder output to the LED segments. The truth table for a seven-segment readout, together with the corresponding BCD inputs, is given in Fig. 5-19. If pulses are present on all segments, but there is no 8 or 9 display, the LED is at fault. As an alternative first step, inject a pulse on all segments of the LED simultaneously, and check for an 8 display.

If the pulses required to make an 8 or 9 display are absent from the LED,

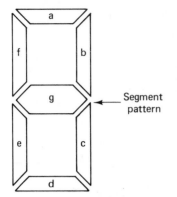

Digit	Segments							BCD inputs			
	a	b	c	d	e	f	g	8	4	2	1
0	1	1	1	1	1	1	0	0	0	0	0
1	0	1	1	0	0	0	0	0	0	0	1
2	1	1	0	1	1	0	1	0	0	1	0
3	1	1	1	1	0	0	1	0	0	1	1
4	0	1	1	0	0	1	1	0	1	0	0
5	1	0	1	1	0	1	1	0	1	0	1
6	0	0	1	1	1	1	1	0	1	1	0
7	1	1	1	0	0	0	0	0	1	1	1
8	1	1	1	1	1	1	1	1	0	0	0
9	1	1	1	0	0	1	1	1	0	0	1

Segment pattern ←

Figure 5-19   Truth table for seven-segment readout pulse and corresponding BCD inputs.

check for an 8 or 9 input to the units decoder. As shown in Fig. 5–19, an 8 input is produced when there is a pulse at the BCD 8-line (between the counter and the decoder). A 9 input to the decoder requires simultaneous pulses on the 8- and 1-lines.

If the 8 and 9 inputs are available to the decoder but there are no 8 and 9 pulses to the LED segments, the units decoder is at fault. As shown in Fig. 5–19, an 8 display requires all segments to receive pulses; a 9 display requires all segments but the *d* and *e* segments. If a logic clip is available, check the operation of the units decoder on a single-stepping basis. That is, inject input pulses at the BCD 8-line (from the counter) while simultaneously enabling the decoder, and check that the outputs to all LED segments go true.

Troubleshooting problem 3.   Assume that the test conditions are the same as for problem 1 except that the input frequency is 300 Hz. However, the readout is 600. That is, the readout is twice the correct value. The logical conclusion here is that the problem is in the counter function rather than in the readout. It is possible that the readout could be at fault, but it is not likely.

A common cause for faults of this type (where FFs are involved) is that one FF is following the input pulses directly. That is, the normal FF function is to go through a complete change of states for two input pulses. Instead, the faulty FF is changing states completely for each input pulse (similar to the operation of a one-shot multivibrator). Thus, the decade counter containing such a faulty FF produces two output pulses to the next counter for every 10 input pulses (or the counter divides by 5 instead of 10). All decades following the defective stage receive two input pulses, whereas they should receive one.

The first practical step is to monitor the input and output of each decade in turn, starting with the units decade. The decade that shows two outputs for 10 inputs, or one output for 5 inputs, is at fault.

Troubleshooting problem 4.   Assume that the test conditions are the same as for problem 1. However, the readout is now 000, 000, 000, 000, and so on. That is, the LEDs glow but remain at 000.

The logical conclusion here is that the LEDs are receiving power and are operative. If not, the LEDs could not produce a 000 indication. The most likely causes for such a symptom are: (1) no input pulses arriving at the units counter; (2) a simultaneous reset pulse with the first input pulse (or a short on the reset line; (3) no enable pulse to the decoder/storage ICs; or (4) a defective units counter (not responding to the first input pulse).

Here, the practical steps are to monitor (simultaneously) the input line and reset line, input line and enable line, and input line and the output of the units counter IC. This should pinpoint the problem. For example, if there are no input pulses (or the wiring is defective, possibly shorted, so that the input pulses never reach the units counter input), there will be no output. If the input pulses are present, but a reset pulse arrives simultaneously with the first input pulse, there can be no readout.

If there is no enable pulse, there will be no readout, even with the counters operating properly. That is, the counters produce the correct BCD output, which is then stored in the decoders. However, the absence of an enable pulse prevents the stored information from being displayed on the LEDs. If the input pulses are present and there is no abnormal reset pulse, the output of the units counter should show one pulse for 10 input pulses. If not, the units counter is defective.

5-3. PROCEDURE WHEN A MICROPROCESSOR DOES NOT YIELD TO ROUTINE TROUBLESHOOTING APPROACHES

The following paragraphs describe a general procedure for resolving problems when a microprocessor-based system does not respond to the conventional troubleshooting procedures described thus far. In brief, the procedure involves breaking the feedback loop into sections. Then each section is "turned on" independently of all other sections. This is sometimes known as "bringing up the microcomputer." Once a section is proved good, you go on to another section until all sections in the feedback loop are operating properly. If each section in the loop is normal, but the system still fails to work, the problem is one of interconnections (unless, of course, the program is simply not practical). In any event, the following procedure will isolate the problem areas. Although the technique described here applies to any microprocessor-based system, it is particularly effective in the development stage.

In this book, we are concerned primarily with troubleshooting rather than debugging. That is, most of the procedures are based on the assumption that the microcomputer or other system was once working properly. However, during the development stage of a microcomputer, the line between trouble-shooting and debugging is difficult to distinguish. This is particularly true for the hobbyist and experimenter. Take the classic example of the hobbyist who has assembled an experimental "microcomputer" similar to that shown in Fig. 5-20. The microprocessor, ROM, RAM, I/O, clock, and reset switches are all interconnected, the keyboard/CRT is connected to the I/O, and power is applied. There is no smoke, arcing, or flame, but nothing seems to work!

Several questions can arise in the hobbyist's mind. Are all of the elements (microprocessor, ROM, RAM, I/O, etc.) good? Is the wiring correct? If the answer is "yes" to both of these, then the problem is in software. The program has bugs. But are you sure?

The basic problem in this situation, and in any microprocessor system where the cause of malfunction is not readily apparent, is one of *feedback*. The microprocessor operates in response to data words appearing on the data bus. In turn, the data words depend on the ROM/RAM address selected by the word on the address bus. Furthermore, the address word depends on the address register in the microprocessor (which depends on the word on the data bus)!

To further confuse matters, the microprocessor probably increments to a

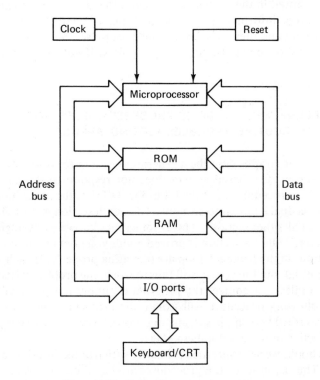

Figure 5-20 Basic microcomputer system.

new address each time the data word is changed. And if there are any words or parts of words appearing on the I/O, these can be applied to the address and data buses. Now imagine if one address line or data line (only one) is open so that it is always at 0 (where one pin of an IC has broken internally). Or consider what occurs if two data lines are interchanged (in the case of a development system). With either of these conditions, the microprocessor delivers an incorrect address to memory (possibly an unused address), and memory plays back an incorrect data byte (possibly a meaningless instruction). Even with a perfect program, the result is chaos. However, the following steps will bring some order to this chaos.

5-3.1 Checking the Clock and Reset Circuits

Once you are sure that power is applied to all the elements in the system, check both the clock and reset circuits before you break into the feedback loop.

The clock line is connected to all the elements (in most systems). Verify this by checking for a stream of good clock pulses at each element, using a logic probe or oscilloscope. The clock pulses must be of correct amplitude (typically about 5 V), width, and frequency. If there are two or more clocks, the clock pulses must be shifted in phase from each other (typically by 90°).

The reset line is usually connected only to the microprocessor. Generally, the reset line is held at 0 V until the reset function occurs. This action applies +5 V to the microprocessor reset pin, and the system is reset. Verify the reset operation by monitoring the address bus, with the reset signal applied. If there is no reset button or control in the system, pulse the reset line, or apply +5 V to the reset line, whichever is most practical. In most systems, the address bus should go to 0000 when the reset line is activated.

5-3.2 Breaking the Microprocessor Loop

The simplest way to isolate the microprocessor is to plug in only the microprocessor, and leave the RAM, ROM, and I/O out of the system. This is generally no problem for development equipment since the ICs are often mounted in sockets. However, there are problems when the system is assembled in final form. Keep in mind that you will use this procedure only when all else has failed, so the inconvenience of disconnecting the ROM/RAM and I/O can be well worth the final results. Leave the clock and reset circuit, as well as all the wiring, in the final form.

Next apply a *no-operation* data word or instruction on the data bus. This instruction may be called by many names (NOP, IDLE, WAIT, etc.) in various microprocessors. However, most microprocessors have some form of no-op instruction. Apply the instruction by connecting each line in the data bus to +5 V or 0 V as needed to form the correct binary word, as shown in Fig. 5-21. Note

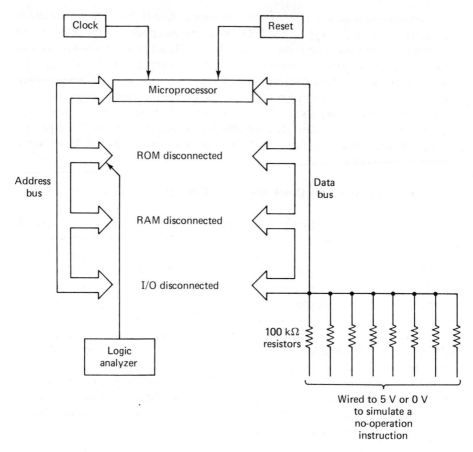

Figure 5–21 Monitoring the address bus with a no-operation instruction applied to the data bus, and the ROM, RAM, and I/O disconnected.

that resistors are used between the supply or ground and the data bus. The resistors are included to prevent damage to the microprocessor. Many microprocessors try to put data onto the bus during an operating cycle. If the data bus is wired directly to the supply or ground, the data output buffer in the microprocessor can be damaged.

An alternative method for setting a no-op word onto the data bus is through three-state gates. Connect the control line for all the gates to the microprocessor read/write line so that the gates are active only when the microprocessor reads data.

With a fixed no-op instruction word on the data line, the program counter in the microprocessor increments repeatedly. That is, the microprocessor executes a no-op, increments the program counter, executes the next no-op, and so on.

5-3.3 Counting Sequence on the Address Line

With the setup shown in Fig. 5-21, the microprocessor should increment continuously through all the addresses. This can be verified by connecting the input probes of an analyzer to all lines of the address bus, and setting the analyzer controls for a count function (most analyzers have some form of count function). Make the connection at the ROM end of the address bus to ensure that correct addresses are being transmitted to the ROM. Start by actuating the reset line and checking for an address or count of 0000. Then return the reset line to 0 V and check that the count goes through FFFF in sequence. In most systems, the microprocessor continues to execute no-ops indefinitely with the Fig. 5-21 setup. That is, when FFFF is reached, the address register continues on to 0000 again, and repeats all addresses.

By using this procedure, you verify that the microprocessor is executing no-ops and that the addresses are correctly transmitted to the ROM. If the addresses do not form a counting sequence, examine the address pattern to determine if address lines are interchanged or are inactive (fixed at 0, 1, or in a high-impedance state), or are shorted, or if the microprocessor is executing an unexpected branch instruction.

5-3.4 Checking Waveforms at Each Change

At this point, it is also important to examine the waveforms on all buses and control lines (read, write, chip enable, etc.) of the system. This can be done with an oscilloscope or with analyzers that have a timing display. All timing functions should be compared against the timing diagrams shown in the microprocessor literature or system service manual. Any incorrect timing, marginal voltage levels, noise, or crosstalk should be eliminated before proceeding.

Timing should be checked as each new block or section is added. Although such testing may seem needlessly repetitive, it takes very little time if there is no problem. If there is a problem, a check at this stage can save a great deal of time. Consider the following problems.

The timing of control signals is particularly important. It is obvious that a control signal (such as a read, write, or chip enable signal) must be applied at exactly the right time. For example, if a data bus buffer is held open too long, all or part of another data word appears on the data bus. Similarly, if the data bus buffer is closed too soon, the data word may not appear in sequence.

Marginal voltage levels can be a problem, particularly on data and address buses. For example, assume that +5 V normally represents a 1, but that a data bus input buffer (say at the microprocessor) recognizes anything above about +3 V as a 1. If the pulses appearing on the data bus drop to something on the order of +2.75 V, they might be recognized as a 0.

Noise and crosstalk can also be a problem on data and address buses, even on control lines. For example, if a line has noise pulses (or crosstalk, which is noise picked up from adjacent lines), and these pulses are near the 2.75-V level, they may appear as 1s. The results are obvious if that particular line is supposed to be at a 0.

5-3.5 Checking the ROM

Once you are satisfied that the microprocessor is functioning normally (executing no-op commands, issuing all control signals such as read/write and chip select, and incrementing through all addresses in sequence), connect the ROM into the circuit. Make all connections as usual, except for the data bus. As shown in Fig. 5-22, the data output of the ROM should be monitored with an

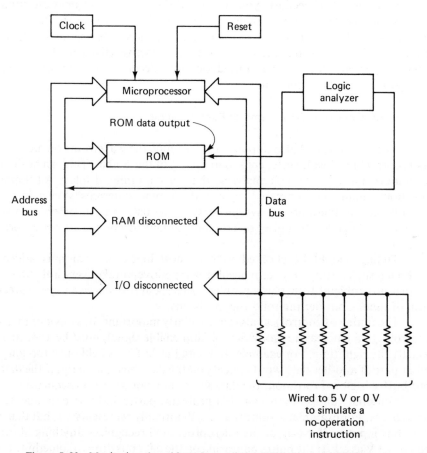

Figure 5-22 Monitoring the address bus and the ROM data output with a no-operation instruction applied to the data bus, the ROM disconnected from the data bus, and the RAM and I/O disconnected.

analyzer, but should not be connected to the data bus. Instead, leave the fixed, no-op instruction word on the data bus (via the resistors or three-state gates). In this way, the microprocessor continues to see a no-op instruction, and continues to increment or count through the addresses as discussed.

This time, however, the ROM cycles through all possible addresses so that you can measure the ROM data outputs with the analyzer. It is assumed that the ROM has some known stored information that you can check on an address-by-address basis. If not, make up a test PROM by programming predetermined words at each address and then inserting the PROM into the ROM socket. (If you are troubleshooting an existing system, the ROM will have its own program. The PROM is used primarily for systems under development.) It is also assumed that if there is more than one ROM, each ROM is checked in turn. If the analyzer has the capacity, continue to monitor all the address lines, as well as the data lines.

Since not all the addresses are ordinarily allocated to the ROM, it might be necessary to connect some temporary pull-ups to the ROM data outputs. This can be done by connecting the lines to +5 V through 100-kΩ resistors. With such pull-ups, an address outside the allocated ROM addresses generates a known data word (all 1s).

Keep in mind that it is not absolutely necessary to monitor all possible ROM outputs, although such a check can be beneficial. However, if there is more than one ROM, you should check sufficiently to verify that the correct ROM is selected, and that every ROM is addressed correctly (in sequence).

You should also check some addresses outside those of the ROM to verify that the ROMs are off when they are not addressed. This is often overlooked. Remember to check the waveforms (Sec. 5-3.4) on the address bus and control lines, particularly on the ROM control lines. It is possible for two ROMs (or a ROM and a RAM) to be on at the same time.

5-3.6 Completing the Microprocessor-to-ROM Link

Once satisified that the ROMs are being addressed in sequence and delivering the correct outputs, connect the ROMs to the data bus and remove the fixed no-op instruction word. In effect, restore the microprocessor-to-ROM link, but do not insert the RAM or I/O at this time.

To make a comprehensive test, install a ROM (or test PROM) with a simple diagnostic routine or program. The test program should contain several unconditional jumps. A flowchart of such a program is shown in Fig. 5-23. Verify operation of the test program by monitoring the address bus with the analyzer.

Note that the program includes RAM access and I/O instructions so that the RAM and I/O control cycles can be checked before the RAM and I/O devices are installed. RAM and I/O timing should also be checked at this point (as discussed in Sec. 5-3.4). Generally, it is not necessary to monitor the data bus unless there is a problem, because the sequence of program addresses is am-

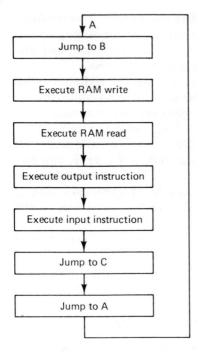

Figure 5-23 Verifying the microprocessor-to-ROM link with a simple diagnostic routine.

ple to verify proper execution of the program. That is, if you have put each of the Fig. 5-23 steps in addresses in sequence, and each address appears in that sequence as monitored on the address bus, the correct word must be appearing on the data bus.

Although the diagnostic program of Fig. 5-23 appears simple, it is generally adequate to test the microprocessor-to-ROM data link, as well as the RAM and I/O control cycles, of a typical microcomputer system. Of course, if another more elaborate program is available in the service literature, use the manufacturer's program. As in the case of other electronic troubleshooting, when all else fails, follow instructions!

In no case should any branches on the RAM or I/O instructions be used at this point, as the RAM and I/O blocks have not yet been turned on and debugged. The checkout of the microprocessor-to-ROM data link is by far the most tedious. The reason is that this link must always be a feedback process. That is, each instruction depends on the address, and each address depends on the previous instruction. The RAM and I/O blocks can be turned on much more directly and in any order. If you choose the RAM first, you can connect it to the system in one operation.

5-3.7 Turning on the RAM

With the RAM connected to the system, run the ROM test program (Fig. 5-23) briefly to verify operation. Pay particular attention to the timing of the

RAM control signals during the RAM read and write instructions. The usual cause of failure at this point is a shorted address or data line, two lines shorted together, or an unwanted RAM response. With the ROM program verified, run a RAM test.

A RAM test program should write to every location in memory, and then read each location back and verify the data. With an 8-bit-wide memory, watch out for the following problem.

The 8 bits of memory represent only 256 states. Conventional memories are usually much longer. This means that each possible data pattern must be written several times to fill the memory. If the same data words are written into each block of 256 words, an error in any of the higher-order addresses can be masked.

An extreme example of such masking is the case where all address lines are disconnected. Any slight change in the 256-word pattern, such as shifting the pattern one word location in each block, will reveal the problem. For example, if you count from 0 to 255 in the first block, you should count from 1 to 255, then go back to 0 in the next block. Next count 2 to 255, and go back to 0 and 1 in the next block, and so on.

Figure 5-24 shows typical connections for testing the RAM functions. Figure 5-25 shows the flowchart of a RAM test program. This test program verifies that the memory system is working correctly, but does not provide a check of each cell at each memory location. However, you can design the program so that all locations are written and then read back. In addition to checking the data words at each address in RAM, check the timing of all RAM functions as discussed in Sec. 5-3.4.

5-3.8 Turning on the I/O

Although the I/O section is relatively easy to turn on, the discussion here is somewhat general, since I/O structures vary more than other blocks from one system to another. The main point is to test the I/O ports before connection to the peripheral devices, such as keyboards, CRTs, or circuits to be controlled.

The first step is to connect the I/O ports into the circuit and then run the ROM test program (Fig. 5-23). This verifies that the control timing is correct with the I/Os in the system. Once you are satisfied that the timing is correct, and that input/output instructions can be executed, the next step is to test the I/O functions with diagnostic programs.

You can check the output ports easily with a simple program that first sets all the ports to 0, then sets each port in turn to 1, and finally sets each port back to 0, one at a time. When testing the output ports, connect the analyzer to one block at a time. If the analyzer has sufficient channels, connect these to the address bus, as well as the input/output ports, as shown in Fig. 5-26. The object

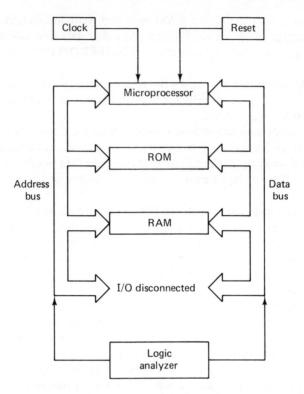

Figure 5-24 Monitoring address and data buses to check RAM functions with a simple diagnostic routine.

of this exercise is to see if the output ports are connected in proper order and can be set to both high and low.

The input ports can be tested in a similar fashion. The program should check for each input high, then for each input low. Figure 5-27 shows a typical input port test program. This program loops until the input under test is forced to the desired state, then jumps to another loop. A simple approach is to pull all the inputs, either high or low through a resistor as shown in Fig. 5-26.

Assume that you select the "high" approach. You then write a program that has two loops (Fig. 5-27). The first loop tests for a specific input low, and the second loop tests for that same input high. While the analyzer monitors the address bus and at least one input under test (preferably all the inputs, if the analyzer has the capability), force the monitored input low with a grounded wire. Then the analyzer shows which loop the microprocessor is in, exactly when the input goes low, and (in a second pass) when the input goes high.

Although this process may seem tedious, the time required to write the test programs must be spent only once. The programs are then available at every phase of system development.

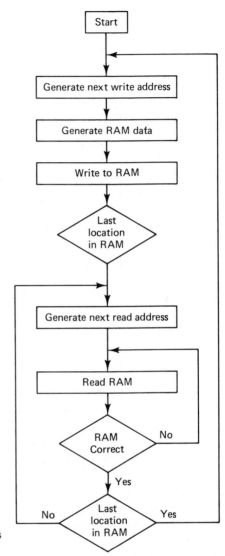

Figure 5-25 Checking RAM functions
with a simple diagnostic routine.

5-3.9 Developing Software During Design and Troubleshooting

To be effective, software must be developed in sections, preferably at the
same time as when the hardware sections are turned on. In this way, you can
develop and test programs in manageable bytes. There are few, if any, wizards
who can write a complete program and have it work the very first time. There
are three approaches to developing software, just as there are for the hardware.
These approaches include single-step, simulator, and analyzer.

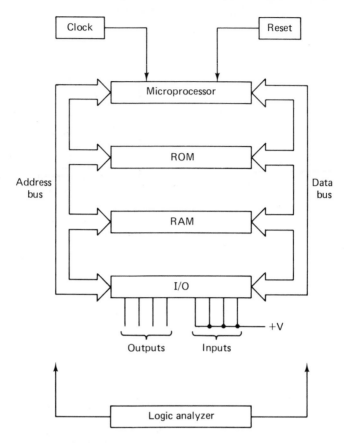

Figure 5-26 Monitoring I/O inputs, outputs, address bus, and data bus to check I/O functions with a simple diagnostic routine.

With any of the approaches, you must write a few steps of program, say to complete one function, and then test that section of program. The obvious drawback to single-stepping is that it is slow and tedious. A not so obvious drawback is that a program may work at the very slow speeds of single-stepping, but not work when the system is operated at normal speeds (1 to 3 MHz). A simulator (either a development system supplied by the microprocessor manufacturer or a large computer or computer time-sharing system) can be a valuable aid in testing complex portions of a program such as mathematical functions. However, it is usually difficult to simulate I/O functions in software, and it is at the I/O ports that major trouble usually develops. It is generally more practical to write an I/O program, and then check the results using actual hardware wired in final form. This is where an analyzer can be of maximum benefit.

When using an analyzer in developing software, either for diagnostic

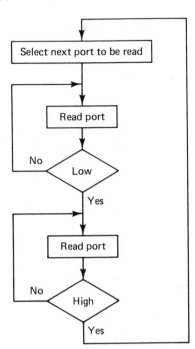

Figure 5-27 Testing I/O input ports with a simple diagnostic routine.

routines or final program functions, it does not really matter whether the software has been simulated beforehand or not. (Of course, if a simulator is available, it can be a help in developing some parts of the hardware, as discussed above.) The main advantage of the analyzer approach is that the hardware and software can be debugged at the same time. Another advantage is that the analyzer can monitor the program flow in single steps, in a group of steps (typically 16 at a time), or can monitor the entire program in a single pass.

In debugging software, you use the analyzer in the same way as when debugging hardware. In fact, most of the hardware debugging techniques described in this section are simply a matter of developing a short program (or taking a section of a larger program), monitoring the flow of the program or section, and then fixing the hardware when the program does not work. (If you are troubleshooting an already developed system, you break the desired program into small sections, and check the flow through the sections.)

In reality, the process of debugging software is really more a problem of identifying the problem, deciding how the software and hardware contribute to the problem, and then doing the fix. The logic analyzer is the ideal tool in pinpointing whether software or hardware is the culprit. Once the hardware is checked out as described in this section (from the clock and reset, through the microprocessor, ROM and RAM, to the I/O), the software can be loaded in small blocks and tried out.

Although you can debug the software in any order, several rules may be helpful. It is usually best to turn on the keyboard or other entry device first, and then turn on any display or output device. That is, check the input/output routines of the program first. Next turn on the remaining hardware and software together.

5-4. ANALYZING MICROCOMPUTER PROGRAM FLOW DURING TROUBLESHOOTING

This section describes how basic microcomputer program flow is analyzed during a typical troubleshooting sequence. These same procedures can also be used during design and development (or debugging) of a microcomputer system. The section concentrates on real-time analysis of program flow using an analyzer. Techniques such as triggering on a specific event and paging of program sections are discussed.

The microprocessor selected for discussion is the M6800 manufactured by Motorola Semiconductor Products, Inc. The M6800, which forms the nucleus of the M6800 microcomputer family, features a 16-bit three-state address bus (allowing 65K addressable bytes of memory) and a three-state 8-bit bidirectional data bus. The M6800 system, which includes both maskable and nonmaskable interrupts, operates at clock rates up to 1 MHz.

The first step in making the analysis is to connect the analyzer probes to the microcomputer system buses and control lines, and then operate the analyzer controls as necessary to produce the desired displays. The exact procedures for such connections and operating control sequence depend on the analyzer. The procedures are found in analyzer user manuals. Here we concentrate on analyzing program flow.

In many cases, it is possible to analyze program flow by monitoring only the address bus. In other cases, it is necessary to monitor both the data and address buses, as well as some of the control lines (such as the read/write and interrupt request). Section 5-4.1 illustrates an example of program flow analysis using nothing but the address bus. Section 5-4.2 illustrates how analysis of the same program flow can be expanded by monitoring both buses and certain control lines.

Keep in mind that all analyzers are not capable of monitoring all these functions, and the program format varies with the type of microprocessor involved. Also, the display format is not the same for all analyzers. However, the same basic procedures apply to analysis of program flow.

5-4.1 Analyzing Program Flow with Address Bus Display Only

Figure 5-28 illustrates system response to an interrupt, and compares the real-time state analysis (as displayed on an analyzer) to the M6800 cross-

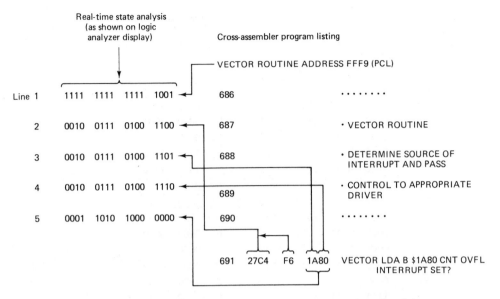

Figure 5-28 System response to an interrupt as shown by comparison of real-time state analysis to cross-assembler program listing.

assembler program listing. The M6800 microprocessor responds to an interrupt in the following sequence.

First, the microprocessor completes the current instruction. Next, the microprocessor stores the contents of its registers on a stack in memory. Next, the program counter is loaded with an interrupt service routine vector fetched from memory locations FFF8 (program counter high, PCH) and FFF9 (program counter low, PCL). Finally, the microprocessor begins execution of the interrupt service routine at the vectored location.

Consider the program listing of Fig. 5-28. Observe the vector routing beginning at location 274C with the instruction LDA B $1A80. This is a three-byte instruction: the first byte (F6) is the operation code, with the section (1A) and third (80) being a double byte operand. In this case, the operand is an extended address.

Proper operation of the vector fetch is confirmed by observing that the address immediately following the trigger word FFF9 is 274C. This means that the microprocessor fetched 27 from FFF8 and 4C from FFF9.

The LDA B $1A80 may be confirmed by observing the second, third, and fourth lines of the state analysis display. Line 2, address 274C, is the fetch of the operation code LDA B (F6), and lines 3 and 4 are the fetch of 1A80. Address 274D contains 1A, and address 274E contains 80. The next line (line 5) shows the address to be 1A80), which implies correct execution of the instruction in the routine. In a similar fashion, each instruction may be shown to have been properly executed.

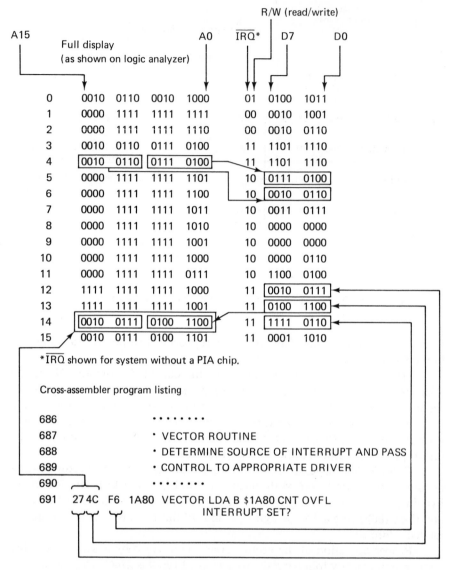

Figure 5-29 System response to an interrupt as shown by comparison of full display (address, data, IRQ and R/W) to cross-assembler program listing.

5-4.2 Analyzing Program Flow with Full Display

Figure 5-29 illustrates an amplification of the example described in Sec. 5-4.1 (analysis of an interrupt response). As shown in Fig. 5-29, when the address and data buses, as well as the control lines (interrupt request IRQ, and read/write) are monitored by the analyzer, it is possible to confirm exact system operation with respect to an interrupt response, including the storing of microprocessor registers on the stack.

With reference to lines 0, 1, and 2 of Fig. 5–29, observe that the IRQ line is low (at 0) for three clock periods. This signals the microprocessor to begin executing an interrupt service operation. Confirmation that the microprocessor completed the current instruction is shown in lines 3 and 4 of Fig. 5–29.

Similarly, it can be shown that the microprocessor saved the internal registers on the stack by an examination of lines 4 through 11. Line 4 shows the completion of the current instruction at location 2674. Line 5 shows the least significant byte of this address (74) being written into location OFFD, while the read/write line is low (0), or in the "write" mode. Line 6 shows the corresponding operation for the most significant byte of the program counter (26) being written into location OFFC. This process continues until all registers are stored on the stack as shown in Fig. 5–30. You can see that seven stack locations are required to save the microprocessor status. This may be confirmed by lines 5 through 11 inclusive of Fig. 5–29.

As discussed in Sec. 5-4.1, the microprocessor then reads memory locations FFF8 and FFF9 to set the interrupt service vector. Observe that the program listing of Fig. 5–29 defines this to be 274C, which means that memory location FFF8 must contain the data 27, and FFF9 must contain 4C.

These functions are confirmed by lines 12 and 13 of Fig. 5–29. Also observe that the fetch was executed properly, as the next address following FFF9 is 274C (line 14). In addition, the data byte at location 274C is F6, which corresponds with the program listing.

5-5. USING AN ANALYZER DURING TROUBLESHOOTING

This section describes use of a typical logic analyzer for several troubleshooting situations. The analyzer chosen for these examples is the Hewlett-Packard 1610A Logic State Analyzer, shown in Fig. 4–15. The instrument is keyboard controlled, provides a tabular display and graph display, and offers general-purpose measurements in microprocessor-based systems, minicomputers, or virtually any digital circuit. The 1610A synchronously performs real-time trace and count measurements up to 10 MHz with powerful triggering capabilities on words up to 32 bits wide, to allow you to capture the data of interest.

Measurements of system activity are displayed on the analyzer's CRT screen in selectable hex, octal, binary, or decimal codes. Setup for measurement is aided with the format and trace specification menus, which indicate the test parameters that you are to enter. (The bottom CRT display of Fig. 4–15 shows a typical menu.) The events and activity that are captured and displayed from the system are gathered at clock transitions after the analyzer locates the specified trace position and then captures 64 words of data.

The displayed trace may be a simple *breakpoint,* with the trace position at

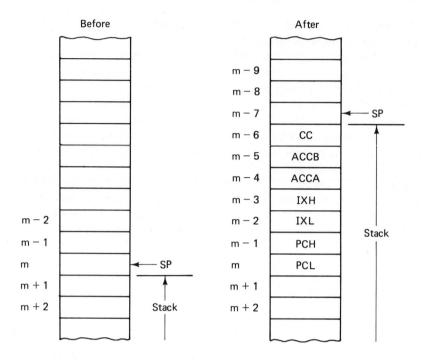

SP, stack pointer
CC, condition codes (also called the processor status byte)
ACCB, accumulator B
ACCA, accumulator A
IXH, index register, higher-order 8 bits
IXL, index register, lower-order 8 bits
PCH, program counter, higher-order 8 bits
PCL, program counter, lower-order 8 bits

Figure 5–30   Saving the microprocessor status during a subroutine using a memory stack.

the beginning, end, or center of the captured data, or in a *state sequence* where one to seven words must be found in a specified order before data bytes are captured. (The center CRT display of Fig. 4–15 shows a typical trace.) The state sequence permits you to locate directly sections of branched, looped, or nested loops of state flow. A selective trace of from one to seven words may be OR-specified, which allows only the words of interest to be captured and eliminates data bytes that are not necessary for your measurement.

A *count measurement* capability allows you to perform a time or state count on all 64 traced states in either absolute or relative modes. With the count measurement you can determine how much time a program spends in loops and servicing interrupts, as well as the time between program steps. This measurement is performed simultaneously with the trace, and all 64 words traced are assigned a count record, which is displayed as positive or negative time in relation to the location of the trace position (absolute mode). In the relative mode, each count recorded is displayed in relation to the previously acquired state.

One complete measurement, including format and trace specifications, may be internally stored to be recalled at a later time or for use in a *trace compare* mode. When a trace compare mode is selected, the display presents an exclusive-OR tabulator listing of the differences between the current and stored measurements. The trace compare mode may also be used to direct the analyzer to continuously rerun a measurement until the current and stored measurements are equal, or not equal, and the analyzer automatically halts and retains the current measurement.

The analyzer includes a *trace graph* to provide a display of data magnitude versus time sequence for all 64 words in memory. Each dot representing a word is given a vertical displacement corresponding to its magnitude and is positioned horizontally in the order of its occurrence. The result is a waveform analogous to oscilloscope displays, and it offers a quick overview of program operation. The trace graph function is, in effect, a form of mapping (although not the form of mapping described in Sec. 4-9.6).

For increased confidence of the instrument's operation, there are self-tests for the keyboard, ROM/RAM, display, a trace test that includes all probe pods, an interrupt test, and a printer test.

Hard copy of both the format and trace specifications, as well as a trace list and trace compare, can be obtained by adding an accessory printer to the basic instrument. Rear panel printer outputs are included in the analyzer for direct interfacing.

Although the following paragraphs provide specific information necessary to set up and use the 1610A (use of controls, display functions, etc.), the troubleshooting situations or problems described here are common to many microprocessor-based systems. When you understand the problems, you can adapt the features of other analyzers to solve the troubleshooting situations. Of course, all the 1610A features are not always available in other analyzers.

5-5.1 Format Specification (Setting Up the Data Inputs)

In this example, the data probe (or *pod,* as it is called) and the clock pod are plugged into the 1610A. When power is applied to the 1610A, the format specification menu is automatically displayed. The menu is similar to that shown in Fig. 4-15, except that all inputs (pods 1, 2, 3, and 4) are labeled F in hex display code. The 1610A is now ready to acquire data without any other setup required.

Pressing the *trace* key on the 1610A results in a *trace list* similar to that in Fig. 5-31. Note that this list has data only in the two right-hand columns. In

Logic analyzer display

```
 — — — — — — TRACE LIST — — — — — — TRACE COMPLETE — — — —

            LABEL        F
            BASE:        HEX

    START            00000062
     +01             00000063
     +02             00000064
     +03             00000065
     +04             00000066
     +05             00000067
     +06             00000068
     +07             00000069
     +08             0000006A
     +09             0000006B
     +10             0000006C
     +11             0000006D
     +12             0000006E
     +13             0000006F
     +14             00000070
     +15             00000071
     +16             00000072
     +17             00000073
     +18             00000074
     +19             00000075
```

Figure 5-31 Trace list showing results of the format specification.

hex, this means activity on all eight channels of pod 1. If you switch back to the format specification menu (Fig. 4–15), you will see exclamation marks on all but the two right-hand digits of pod 1.

There are now many ways to set up the input channels in the format specification. For example, all unused inputs can be turned off (so they are not displayed) by positioning the cursor in the appropriate pod display and entering "don't cares" as indicated by an X) in all pods except pod 1. Since pod 1 is connected to a two-unit hex counter, it might be desirable to display each unit separately by assigning a different label to each counter. In this example, arbitrarily assign label A to the upper unit counter and label C to the lower unit counter. Under these conditions, pressing the trace key provides a trace list, as shown in Fig. 5–32, with the display listed by label and in hex format.

If desired, the numerical base for label C can be changed to binary. The resultant trace list is then as shown in Fig. 5–33. Any combination of labels or numerical base selection can be assigned to the input channels. However, if labels are not assigned sequentially, but are mixed, an illogical condition results

Logic analyzer display

```
┌──────────────────────────────────────────────────────────────────┐
│  — — — — — — TRACE LIST — — — — — TRACE COMPLETE — — — — —          │
│                                                                    │
│          LABEL     A       C                                       │
│          BASE:     HEX     HEX                                     │
│                                                                    │
│      START         F       2                                       │
│       +01          F       3                                       │
│       +02          F       4                                       │
│       +03          F       5                                       │
│       +04          F       6                                       │
│       +05          F       7                                       │
│       +06          F       8                                       │
│       +07          F       9                                       │
│       +08          F       A                                       │
│       +09          F       B                                       │
│       +10          F       C                                       │
│       +11          F       D                                       │
│       +12          F       E                                       │
│       +13          F       F                                       │
│       +14          0       0                                       │
│       +15          0       1                                       │
│       +16          0       2                                       │
│       +17          0       3                                       │
│       +18          0       4                                       │
│       +19          0       5                                       │
└──────────────────────────────────────────────────────────────────┘
```

Figure 5-32   Trace list with display listed by label and in hex format.

and the 1610A requests a correction of the error (such as a note "WARN-ING SPLIT LABEL" on the display).

5-5.2 Trace Specification

The trace specification menu, which is called to the display with the trace specification key, permits the user to define what data bytes are acquired as well as how they are acquired. Notice that the label base at the top of the display (Figs. 5-31, 5-32, and 5-33) is as defined from the format specification.

The next function to enter is the *trigger conditions* for the trace. The conditions are entered by positioning the cursor in the desired label and entering the trigger word. A trigger value may be entered for each field selected in the format specification. The trigger condition may be positioned to start a trace, be in the center of a trace, or end a trace. For even more trace conditioning, the *insert* and *delete* edit keys may be used to specify up to seven levels of sequential triggering. In addition, each level of triggering may be specified to occur a multiple

Logic analyzer display

LABEL	A	C
BASE:	HEX	BIN

— — — — — — TRACE LIST — — — — — — TRACE COMPLETE — — — — —

LABEL BASE:	A HEX	C BIN
START	F	0010
+01	F	0011
+02	F	0100
+03	F	0101
+04	F	0110
+05	F	0111
+06	F	1000
+07	F	1001
+08	F	1010
+09	F	1011
+10	F	1100
+11	F	1101
+12	F	1110
+13	F	1111
+14	0	0000
+15	0	0001
+16	0	0010
+17	0	0011
+18	0	0100
+19	0	0101

Figure 5-33   Trace list with split label (A in hex, C in binary).

number of times (1 to 65,536) before its sequential condition is satisfied. The following explanations demonstrate the ability of the 1610A to trace simple as well as complex program flow.

Trace sequence.   For simple in-line program flow such as that shown in Fig. 5–34, the trace specification is set to detect state 2848 and then trace the program flow until memory is full (63 remaining states). Now, if a branch (Fig. 5–35) is added around the in-line flow of Fig. 5–34, sequential conditions must be added to the trace specification to enable tracing of a specific branch. As shown in Fig. 5–35, the 1610A must first recognize state 2848, then 287F, and the trace list starts at state 284D.

Unfortunately for the troubleshooter, programs rarely are as simple as those shown in Figs. 5–34 and 5–35. The program in Fig. 5–36 is somewhat more typical of even a simple microprocessor-based system. Figure 5–36 also shows the trace specification for following a specific path (given as path 2) through the network.

In addition to being able to follow a particular branch of a network, the

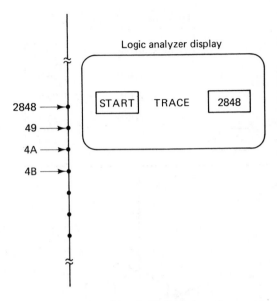

Figure 5–34 Trace specification with a single-state trigger for acquiring in-line program flow.

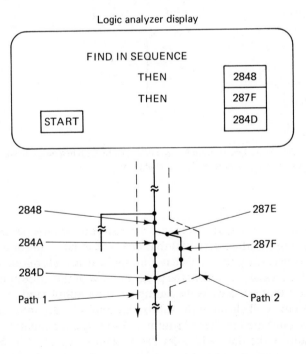

Figure 5–35 Trace specification for tracing a specific branch of program flow.

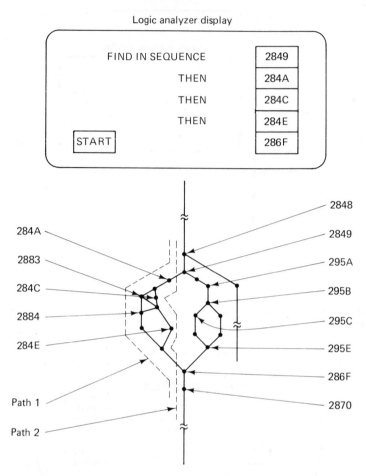

Figure 5-36   Trace specification for tracing a specific path in a program with a complex (but commonly encountered) branching network.

sequence capability of the 1610A makes it possible to acquire parameters (such as data values at specific addresses) that are used in later subroutines. For example, if it is necessary to check that a multiplication subroutine is operating properly, it is necessary to look at the program flow of the subroutine, as well as the parameters that are used as the multiplier and multiplicand. Normally, the parameter occurs well ahead of the subroutine and is loaded into registers to be recalled later (and used in the subroutine). The sequence capability, shown in Fig. 5-37, allows the data addresses 286A, 286D, and 2877 to be acquired. When the trace starts at 29A7, the parameters are displayed in the trace list just prior to the trace listing, as shown in Fig. 5-38.

Logic analyzer display

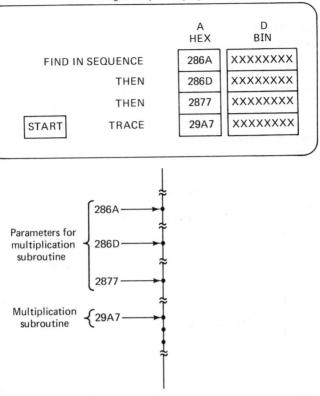

Figure 5-37 Trace specification demonstrating how the 1610A is set to acquire parameters used in a subsequent program subroutine.

Logic analyzer display

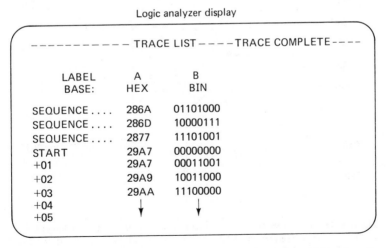

Figure 5-38 Trace list showing the parameters in the D label used in a subroutine starting at 29A7.

Sequence restart.   In addition to being able to follow a specific path in a complex program network, it is often necessary to analyze program flow only if the path is completed in one pass. In some complex networks, such as that shown in Fig. 5-39, it is possible to first follow the sequential conditions of path 1 (satisfying states 2849 and 284A) and then exit the network. This means that the 1610A is looking for the next sequential condition 284C. When the program returns to this network and follows path 3, the remaining conditions are satisfied, and the 1610A captures the data. However, the resultant trace list, beginning at state 286F, does not show the correct activity for a path 2 branch.

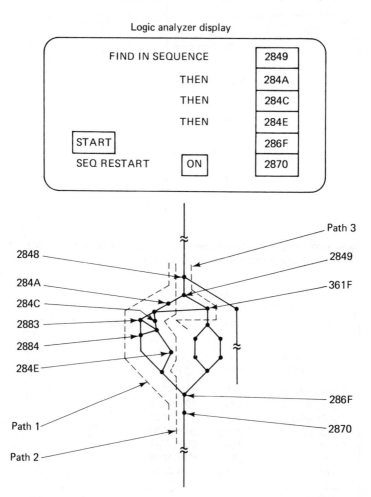

Figure 5-39   Trace specification for tracing a specific path in a branching network, and using sequence restart on an exit state to ensure that the specific path is satisfied after only one pass through the network.

By turning on the *sequence restart* function of the analyzer, and specifying an exit (or entry) state of the network, the 1610A monitors both the sequential condition and the restart condition. This examination is done simultaneously on each state acquired. If the sequence condition is satisfied, the analyzer advances to the next state, disregarding whether restart is satisfied or not. If the sequence is not satisfied and the restart is satisfied, a restart is executed on the entire sequence.

As an example, if path 1 (Fig. 5–39) is followed initially, states 2849 and 284A are satisfied. However, path 1 leads to exit state 2870, which recycles to the first sequence state of 2849. Note that a sequence restart state at the entry of the network at state 2848 will work just as well as the exit state for a sequence restart.

Another major difficulty found in tracing program flow is to trace a *direct jump* (sometimes called a *zero path branch*). The three common forms of program paths—*variable, fixed,* and *zero path branch*—are shown in Fig. 5–40. Defining a trace on branch 1 (variable) or branch 3 (fixed) is easy with a sequential state trace specification. However, the zero length (branch 2) cannot be defined reliably by specifying "find in sequence 2849, start 287C," because either branch 1 or branch 3 also satisfies these conditions.

The solution for capturing a zero length branch is to use a sequence restart with a restart state of don't cares (all Xs in the case of the 1610A). With a don't care restart state, the 1610A first looks to see if a sequence condition is satisfied and, if not, examines if the restart condition is satisfied. Now, if branch 1 or 3 is followed, a state after 2849 other than 287C is found, the restart condition (don't care) is satisfied, and the 1610A returns and looks for 2849. Only when the sequence of 2849, directly followed by 287C, is found will a trace be captured.

Multiple occurrences.   Another form of program flow is the *nested loop,* in which a set of states occurs a number of times prior to advancing the program. These subloops within loops are referred to as *nesting.* An example of programs requiring nested loops is a microcomputer that progressively samples many peripherals (a major loop), receives data from each peripheral (a minor loop), and then formats the data from each data transmission (a subminor loop). The need for tracing loops is common where it is necessary to examine program activity only when it is working on the nth pass of a subminor loop, such as examining the sixth data format of the fourth data transmission of peripheral 3.

The 1610A is capable of determining the number of occurrences of each state sequence. This allows nested loop routines to be traced. The three-level nested loop in Fig. 5–41 has a major loop (loop I) that repeats (or iterates) 17 times, a minor loop (loop J) that iterates 11 times for each occurrence of the I loop, and a subminor loop (loop K) that iterates 13 times for each occurrence of

Logic analyzer display

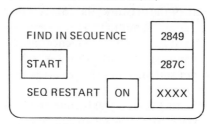

Branch 1: Variable path length
Branch 2: "Zero" path length
Branch 3: Fixed path length

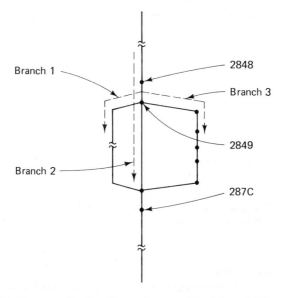

Figure 5-40 Trace specifications for tracing a zero-length branch (direct jump) using sequence restart.

the J loop and 143 times for each occurrence of the I loop. A trace specification can be set in that directs the 1610A to acquire data at state 2841 only when the program is in the ninth pass of the I loop, the eighth pass of the J loop, and the seventh pass of the K loop. Note that the occurrence counter for each state is represented as a decimal number, for ease of use. The range of each counter is from 1 to 65,535.

5-5.3 Digital Delay

Many logic analyzers provide digital delay for the examination of data at some point that is at a number of clock pulses from a reference state or a trigger point. As shown in Fig. 5-40, the digital delay capability is useful only for a

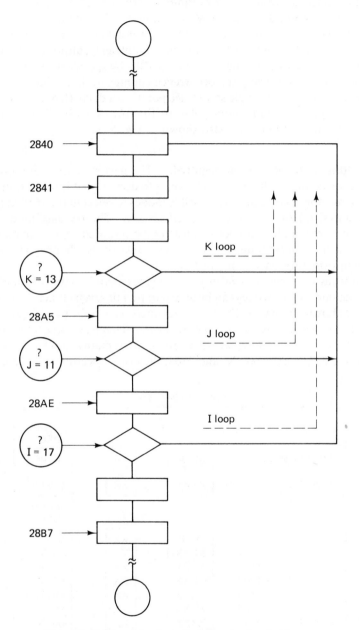

Figure 5–41 Example of a typical three-level nested loop.

path of fixed length (branch 3, for example). Trying to examine a variable path length using a preset delay causes variation in acquired data.

However, if proper sequence states are not known to define a specific path of fixed length, digital delay is useful. Digital delay, although not specified directly, can be easily implemented on the 1610A by specifying a sequence state of all don't cares and using the occurrence counter to determine the delay. A trace specification for a single state trace point with delay (from 1 to 65,536) is shown in Fig. 5-42. Maximum delay in the 1610A is 393,216, and is implemented with six sequences, also shown in Fig. 5-42.

Setting up the data to be acquired.   The next step after defining where information is traced (the trace sequence) is to decide what information is to be stored in memory. There are two possible modes to select in the 1610A: (1) tracing all states, (2) or tracing only specific states. The *trace-all* mode is self-explanatory in that all states occurring after the specified trigger conditions are stored. The *trace-only* mode allows the user to selectively edit program flow and acquire information only of interest.

By using the *insert* and *delete* key when the cursor is in the trace-only field, additional lines (up to seven) can be added to permit seven choices of qualifications, as shown in Fig. 5-43. Note that there can only be a total of eight specifications for trace-sequence and trace-only. That is, if three trace-sequence conditions are defined, only five trace-only states remain available. If the states selected are 2842, 0001, 60XX, and 2864, then only specific states 2842, 0001,

Logic analyzer display

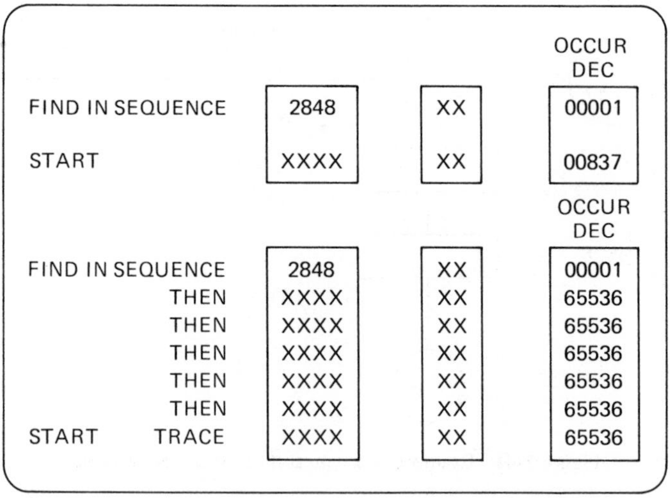

			OCCUR DEC
FIND IN SEQUENCE	2848	XX	00001
START	XXXX	XX	00837
			OCCUR DEC
FIND IN SEQUENCE	2848	XX	00001
THEN	XXXX	XX	65536
THEN	XXXX	XX	65536
THEN	XXXX	XX	65536
THEN	XXXX	XX	65536
THEN	XXXX	XX	65536
START TRACE	XXXX	XX	65536

Figure 5-42   Digital delay implemented by specifying "don't cares" as a sequence condition, and using the occurrence counter for delaying the window.

Logic analyzer display

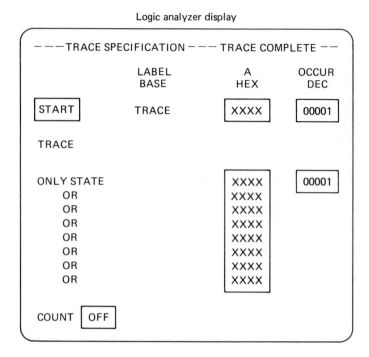

Figure 5-43   Display showing trace-only field, which permits seven choices of quali-
fication for acquiring only information of interest.

2864, and anything within the range from 6000 to 60FF are stored in memory.
Figure 5-44 is the resultant trace list for this trace specification.

To reduce the acquired information further, as well as to view data bytes
that are not of interest (and thus eliminate the need for a large memory in the
analyzer), a multiple-occurrence counter on trace-only is available. The effect
of this counter is to permit periodic acquisition of the specified states only after
they have occurred the number of times set in the counter. For example, if it is
specified to trace only state 2842, with an occurrence of eight, then only every
eighth occurrence of state 2842 is stored, thereby increasing the memory capacity by a factor of 8.

5-5.4 Count Measurements

Microprocessor-based devices are often compared on the basis of quan-
titative evaluations, such as (1) the *elapsed time interval* for two different
devices to execute the same programs, (2) the time interval for one device to execute two different programs (to measure program efficiency), or (3) counting
events during execution of specific programs (such as the number of byte
transfers during a direct memory access, or DMA). These functional
measurements give users the insight as to which system to buy, how to optimize

Logic analyzer display

```
 ┌──────────────────────────────────────────────────────────────┐
 │  — — — — — — TRACE LIST — — — — — — TRACE COMPLETE — — — —    │
 │                                                                │
 │         LABEL        A          D                              │
 │         BASE:       HEX        HEX                             │
 │                                                                │
 │      SEQUENCE       28B7        86                             │
 │      SEQUENCE       28AE        86                             │
 │      SEQUENCE       28A5        86                             │
 │       START         2841        00                             │
 │        +01          2842        9B                             │
 │        +02          0001        03                             │
 │        +03          2842        9B                             │
 │        +04          0001        03                             │
 │        +05          2842        9B                             │
 │        +06          0001        03                             │
 │        +07          60F0        FF                             │
 │        +08          60F1        FF                             │
 │        +09          60F2        FF                             │
 │        +10          2864        60                             │
 │        +11          60F4        FF                             │
 │        +12          60F5        FF                             │
 │        +13          60F6        FF                             │
 │        +14          60F6        FF                             │
 │        +15          60F7        FF                             │
 │        +16          60F8        FF                             │
 └──────────────────────────────────────────────────────────────┘
```

Figure 5-44   Trace list for selected states 2842, 0001, 60XX, and 2864.

programs by detecting slow, inefficient subroutines, and, in general, where problems exist in systems.

The 1610A has two modes of displaying time interval and state count measurements, the *absolute* and *relative* modes. The difference between the two modes is simply the choice of one reference point. In the absolute mode, the count is between the start (center, or end) of the trace, and the acquired state (Fig. 5-45). In the relative mode, the count is between each state and the previously acquired state (Fig. 5-46). These two modes allow maximum flexibility for making evaluations such as benchmarks (Sec. 5-5.5) or optimizing programs, and detecting errors.

5-5.5 Benchmark Using an Analyzer

Performing a benchmark that represents a valid figure of merit for a microprocessor-based system is generally quite difficult. Such considerations as word size, functional interest of the architecture, I/O techniques, and so on, all contribute to the total evaluation of a system. The elapsed time interval for the

Logic analyzer display

```
————— TRACE LIST ———————— TRACE COMPLETE ————
```

LABEL BASE:	A HEX	D HEX	TIME DEC
SEQUENCE	28B7	86 −	1.183 S
SEQUENCE	28AE	86 −	94.05 MS
SEQUENCE	28A5	86 −	6.049 MS
START	2841	00	0 US
+01	2842	9B +	4.0 US
+02	0001	03 +	8.0 US
+03	2842	9B +	481.9 US
+04	0001	03 +	485.9 US
+05	2842	9B +	959.9 US
+06	0001	03 +	963.9 US
+07	60F0	FF +	1.079 MS
+08	60F1	FF +	1.087 MS

Figure 5-45 Time interval displayed in the absolute mode shows the time between trace position and the acquired trace.

Logic analyzer display

```
——————— TRACE LIST ————— TRACE COMPLETE —— ——
```

LABEL BASE:	A HEX	D HEX	TIME DEC
SEQUENCE	28B7	86	
SEQUENCE	28AE	86	1.089 S
SEQUENCE	28A5	86	00.00 MS
START	2841	00	6.049 MS
+01	2842	9B	4.0 US
+02	0001	03	4.0 US
+03	2842	9B	473.9 US
+04	0001	03	4.0 US
+05	2842	9B	474.0 US
+06	0001	03	4.0 US
+07	60F0	FF	116.0 US
+08	60F1	FF	8.0 US

Figure 5-46 Time interval in the relative mode shows the time between each state and the previously acquired state.

execution of a specific task is one of the important considerations. Figure 5-47 shows a setup for a trace specification to measure the time intervals of the sample program, which is composed of four segments.

Quite often, it is important to know not only the total program time, but also the time of each segment. The resultant trace list of Fig. 5-48 indicates that the total program time of 2.168 s between 617E (the beginning) and A83C (the end of the program). Since this is the absolute mode, all states occurring prior to the start are shown as minus (−). Similarly, the beginning time of all the segments is shown referenced to the end of the program (or start of the trace).

Figure 5-49 gives the time interval in the relative mode. (The count measurement is done simultaneously with the trace of data. Selecting either absolute or relative merely changes the output display and does not require rerunning the measurement.) In the relative mode, the individual elapsed time of each segment is displayed. Here, the interval is between each state and the previous state.

5-5.6 Detecting Program Errors with Elapsed Time Comparisons

Being able to analyze the amount of time spent between two successive states in a microprocessor-based system can help in rapidly spotting errors or conditions leading to errors. Examination of Fig. 5-46 shows some interesting activity at states +03, +05, and +07. The elapsed time between these states and

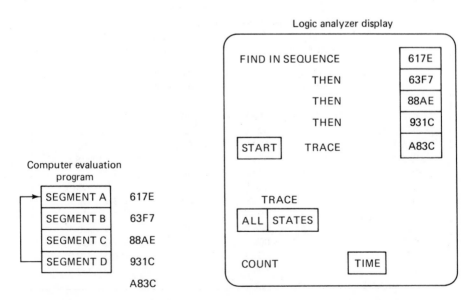

Figure 5-47 Trace specification for a benchmark test that evaluates the time for two systems to execute the same program.

Logic analyzer display

```
 ————————— TRACE LIST ——————TRACE COMPLETE————

                  LABEL      A           TIME
                  BASE:      HEX         DEC

                                      ┌─────┐
                                      │ ABS │
                                      └─────┘
   SEQUENCE                  617E −      2.168 S
   SEQUENCE                  63F7 −     75.89  MS
   SEQUENCE                  88AE        7.395 MS
   SEQUENCE                  931C −     87.0   US
    START                    A83C        0.0
    +01                      ────       ────
    +02                      ────       ────
    +03                      ────       ────
```

Figure 5-48   Trace list of a benchmark showing a time interval measurement of specific states referenced to the start of the trace (absolute mode).

the previous state (which all have address 0001 and data 03) is 473.9, 474.0, and 116.0 μs, respectively. The question can immediately be asked: Why does it take so much more time to go from state 0001 to 2842 (474.0 μs) than it takes to go from 0001 to 60F0 (116.0 μs)? Of course, a detailed examination of the program listing is required to answer this question specifically, but it requires only a brief look at relative elapsed time to locate the apparent problem.

Logic analyzer display

```
 ——————————TRACE LIST —————— TRACE COMPLETE-————

                  LABEL      A           TIME
                  BASE:      HEX         DEC

                                          REL
   SEQUENCE                  617E
   SEQUENCE                  63F7        2.092 S
   SEQUENCE                  88AE       68.53  MS
   SEQUENCE                  931C        7.272 MS
    START                    A83C       87.0   US
    +01                      ────       ────
    +02                      ────       ────
    +03                      ────       ────
```

Figure 5-49   Trace list of a benchmark showing a time interval measurement of specific program segments. The interval is referenced to the previously acquired state (relative mode).

5-5.7 Verification of Program Activity Using a State Count

Once a program has has been written to perform a task, the next step is to run the program and verify that the program actually does the task. It is sometimes possible to be fooled by a program. For example, a program may eventually perform its task, but it may also perform other nondesired activities. By analyzing the program states using a state count in both the absolute and relative modes, the programmer can easily verify the program.

An example of this is found in the nested loop program of Fig. 5–41. To verify that the program actually executes the number of loops specified, count the number of occurrences of state 2840, and compare this count to what is programmed. Figure 5–50 shows the calculation of how many times a state (2840) of the innermost loop should occur before the trace (as specified) begins. The resultant trace list of a relative count of state 2840 is shown in Fig. 5–51. As shown, the count is exactly as calculated. That is, 1287 counts of state 2840 occur between 28B7 and the ninth occurrence of 28AE, and 104 counts of 2840 occur between the ninth occurrence of 28AE and the eighth occurrence of 28A5, and so on.

However, Fig. 5–52 shows the same data displayed in the absolute mode. Here, the number of states 2840 between 28B7 and 2841 (the total before the trace starts) is shown to be 1398 (as calculated in Fig. 5–50).

5-5.8 Comparing Dynamic Data Versus Stored Data

The trace compare function is useful in detecting differences or similarities among data. Quite often, it is necessary to monitor for intermittent state changes, or to compare the contents of one memory unit relative to another.

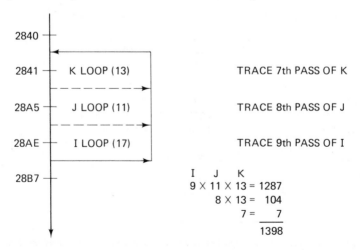

Figure 5-50 Calculations of how many times state 2840 of the innermost loop should occur before the specified trace should begin.

Logic analyzer display

```
 ─ ─ ─ ─ ─ ─ ─ TRACE LIST ─ ─ ─ ─ ─ ─ TRACE COMPLETE ─ ─ ─ ─

        LABEL      A         D       STATE COUNT
        BASE:      HEX       HEX         DEC

  SEQUENCE        28B7      86
  SEQUENCE        28AE      86          1287
  SEQUENCE        28A5      86          104
    START         2841      00           7
    +01           0000      00           0
    +02           2842      9B           0
    +03           2843      01           0
```

Figure 5-51   Resultant trace list of a relative count of state 2840.

The 1610A can be used to monitor incoming data constantly and compare them with known stored data. When a nonequality or equality (of all 32 by 64 bits) is detected, the acquisition of data stops, and the difference is indicated in a listed display.

An example of a possible situation is indicated in Fig. 5-53. The format specification shows two labeled data fields, A and D, both in a display code of binary. A trace specification is then defined, and information is acquired by executing a trace. All 2048 bits of acquired data are then placed in memory A. These data bits are transferred to an internal memory by pressing the *store* key. Both memories then contain the same information. By pressing the *compare versus store* key, a list of the exclusive-OR results between A and D memories is shown. Any differences are indicated and decoded according to the display code selected for the fields.

Logic analyzer display

```
 ─ ─ ─ ─ ─ ─ ─ TRACE LIST ─ ─ ─ ─ ─ ─ TRACE COMPLETE ─ ─ ─ ─

        LABEL      A         D       STATE COUNT
        BASE:      HEX       HEX         DEC

  SEQUENCE        28B7      86          1398
  SEQUENCE        28AE      86          111
  SEQUENCE        28A5      86           7
    START         2841      00           0
    +01           0000      00           0
    +02           2842      9B           0
    +03           2843      01           0
```

Figure 5-52   Resultant trace list of an absolute count of state 2840.

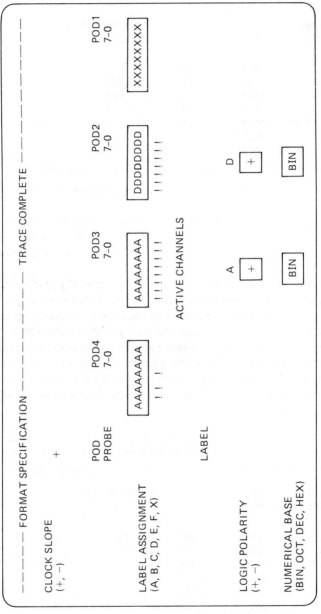

Figure 5-53 Format specification of two labeled data fields for comparing dynamic versus stored data.

Logic analyzer display

```
┌──────────────────────────────────────────────────────────────────────┐
│  ─────── TRACE COMPARE ─── COMPARED TRACE COMPLETE ──                   │
│                                                                        │
│         LABEL      A              D            COMPARED                 │
│         BASE:      BIN            BIN          TRACE MODE               │
│                                                                        │
│        +18    0000000000000000   00000000         0                    │
│        +19    0000000000000000   00000000                              │
│        +20    0000000000000000   00000000                              │
│        +21    0000000000000000   00000000                              │
│        +22    0000000000000000   00000000                              │
│        +23    0000000000000000   00000000                              │
│        +24    0000000000000000   00000000                              │
│        +25    0000000000000000   00000101                              │
│        +26    0000000000000000   00000000                              │
│        +27    0000000000000000   00000000                              │
│                                                                        │
└──────────────────────────────────────────────────────────────────────┘
```

Figure 5-54   Results of a trace compare shows a mismatch in position + 25.

The trace compare menu is a selectable field that allows the acquisition of data to stop when the comparison with stored data is either equal (=) or not equal (≠). In addition, the stopping of acquired data can be turned off so that the 1610A can free-run while indicating the dynamic mismatch of incoming versus stored data.

The results of a trace compare when the *stop when not equal* mode is selected is shown in Fig. 5-54. In this example, the 1610A is continuously comparing the incoming data with stored data. When a mismatch is found (position + 25 in the D label), the acquisition of data stops, and the difference is indicated by 1s in the appropriate channels.

6

NTSC Signals
Used in Television
and VCR Troubleshooting

It is assumed that you are familiar with the basics of television and VCR (video cassette recorder) circuits and troubleshooting, at least at a level found in the author's best-selling *Handbook of Simplified Television Service* (Englewood Cliffs, N.J.: Prentice-Hall, Inc., 1977) and *Complete Guide to Videocassette Recorder Operation and Servicing* (Englewood Cliffs, N.J.: Prentice-Hall, Inc., 1983). The information in this chapter is based on the troubleshooting methods covered in those books and in Chapter 1 of this book, However, as discussed in the two books listed above, troubleshooting of television and related circuits can be improved greatly by the use of an NTSC color bar generator. The signals produced by such generators simulate those broadcast by television stations, and are absolutely essential for proper service and troubleshooting of VCRs. For that reason, this entire chapter is devoted to the NTSC color bar generator, and to the signals produced by these generators.

6-1. THE NTSC COLOR VIDEO SIGNAL

Before going into the NTSC color bar generator, let us review the nature of the NTSC color video signal.

6-1.1 History of the NTSC Color Video Signal

In 1953, the NTSC (National Television Systems Committee) established the color television standards now in use by the television broadcast industry in the United States and many other countries. The color system was compatible with the monochrome (black and white) system that previously existed. The makeup of a composite video signal is dictated by NTSC specifications. These specifications include a 525-line interlaced scan, operating at a horizontal scan frequency of 15,734.26 Hz and a vertical scan frequency of 59.94 Hz. Color information is contained in a 3.579545-MHz subcarrier. The phase angle of the subcarrier represents the hue; the amplitude of the carrier represents the saturation.

6-1.2 Horizontal Sync

A line of horizontal scan is normally thought of as "beginning" at the leading edge of the horizontal blanking pedestal (see Fig. 6-1). In a television receiver, the horizontal blanking pedestal starts as the electron beam of the CRT reaches the extreme right-hand edge of the screen (plus a little extra in

		Horizontal sync pulse	Burst	Gray (75% white)	Yellow	Cyan	Green	Magenta	Red	Blue	Black
Luminance	0	−40	0	77	69	56	48	36	28	15	7.5
Chroma amplitude	−	−	40	−	62	88	82	82	88	62	−
Vector (from B−Y)	−	−	180°	−	167°	284°	241°	61°	104°	347°	−

IEEE units →

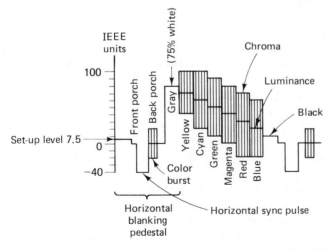

Figure 6-1   Composite video signal showing one horizontal line of NTSC color bar signal.

most cases). The horizontal blanking pedestal prevents illumination of the screen during retrace (until the electron beam deflection circuits are reset to the left edge of the screen and ready to start another line of video display). The entire horizontal blanking pedestal is at the blanking level or the sync pulse level. In a television receiver, the blanking and sync pulse levels are the "blacker than black" levels that assure no illumination during retrace.

The horizontal blanking pedestal consists of three discrete parts: the front porch, the horizontal sync pulse, and the back porch. The front porch is a 1.47-μs period at blanking level. The front porch is followed by a 4.89-μs horizontal sync pulse at the -40 IEEE units level. (An explanation of IEEE units is given in Sec. 6–1.4.) When the horizontal sync pulse is detected in a television receiver, flyback is initiated, rapidly ending the horizontal scan and rapidly resetting the horizontal deflection circuit for the next line of horizontal scan. The horizontal sync pulse is followed by a 4.40-μs back porch at the blanking level. When a color signal is being generated, 8 to 10 cycles of 3.579545-MHz color burst occurs during the back porch. The color burst signal is at a specific reference phase. In a color television receiver, the color oscillator is phase locked to the color burst reference phase before starting each horizontal line of video display. When a monochrome signal is being generated, there is no color burst during the back porch.

6–1.3 Vertical Sync

A complete video image as seen on a television screen is called a *frame*. A frame consists of two interlaced vertical fields of 262.5 lines each. The image is scanned twice at a 60-Hz rate (59.94 Hz, to be more precise), and the lines of field 2 are offset to fall between the lines of field 1 (the fields are interlaced) to create a frame of 525 lines at a 30-Hz repetition rate.

At the beginning of each vertical field, a period equal to several horizontal lines is used for the vertical blanking interval (see Fig. 6–2). In a television receiver, the vertical blanking interval prevents illumination of the CRT during the vertical retrace. The vertical sync pulse, which is within the vertical blanking interval, initiates reset of the vertical deflection circuit so that the electron beam returns to the top of the screen before video scan resumes. The vertical blanking interval begins with the first equalizing pulse, which consists of six pulses one-half the width of horizontal sync pulses but at twice the repetition rate. The vertical sync pulse occurs immediately after the first equalizing pulse. The vertical sync pulse is an inverted equalizing pulse with the wide portion of the pulse at the -40 IEEE units level, and the narrow portion of the pulse at the blanking level. A second equalizing pulse occurs after the vertical sync pulse, which is then followed by 13 lines of blanking level (no video) and horizontal sync pulses to assure adequate vertical retrace time before resuming video scan. The color burst signal is present after the second equalizing pulse.

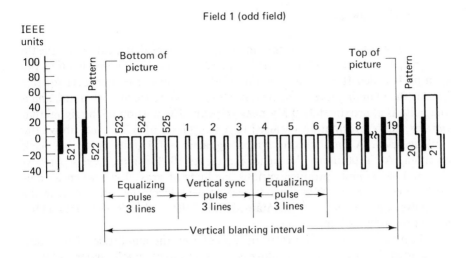

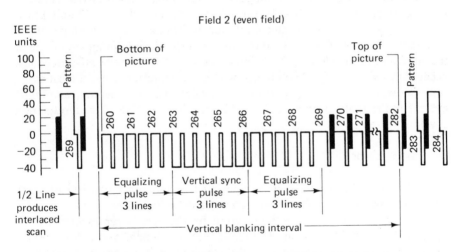

Figure 6-2 Composite video signal showing vertical blanking interval.

Note that in field 1, line 522 includes a full line of video, while in field 2 line 260 contains only a half-line of video. This timing relationship produces the interlace of fields 1 and 2. The patterns produced by an NTSC color bar generator are interlaced. This interlaced scan produces a slight vertical jitter in convergence patterns (available with most NTSC bar generators), but in no way degrades the ability to perform convergence adjustments. Some low-cost color bar/pattern generators produce more stable convergence patterns because they use progressive scan rather than interlaced scan. However, such signals are not to NTSC standards.

6-1.4 Amplitude

A standard NTSC composite video signal is 1 V peak to peak, from the tip of a sync pulse to 100% white. This 1-V peak-to-peak signal is divided into 140 equal parts called IEEE units (see Fig. 6-1). The zero reference level for this signal is the blanking level. The tips of the sync pulses are at -40 units, and a sync pulse is approximately 0.3 V peak to peak. The portion of the signal that contains video information is raised to a setup level of $+7.5$ units above the blanking level. A monochrome video signal at $+7.5$ units is at the black threshold. At $+100$ units, the signal represents 100% white. Levels between $+7.5$ and $+100$ units produce various shadings of gray. Even when a composite video signal is not at the 1-V peak-to-peak level, the ratio between the sync pulse and video must be maintained: 0.3 of the total for sync pulse and 0.7 of total for 100% white.

There is also a specific relationship between the amplitude of the composite video signal and the percentage of modulation of an RF carrier. A television signal uses negative modulation, where the sync pulses (-40 units) produce the maximum peak-to-peak amplitude of the modulation envelope (100% modulation) and white video ($+100$ units) produces the minimum amplitude of the modulation envelope (12.5% modulation). This is very advantageous, because the weakest signal condition, where noise interference can most easily cause snow, is also the white portion of the video. There is adequate amplitude guard band so that a peak white of $+100$ units does not reduce the modulation envelope to zero.

6-1.5 Color

The color information in a composite video signal consists of three elements: luminance, hue, and saturation (refer to Fig. 6-3).

Luminance, or brightness perceived by the eye, is represented by the amplitude of the video signal. The luminance component of a color signal is also used in monochrome receivers, where the luminance is converted to a shade of gray. Yellow is a bright color and has a high level of luminance (is nearer to white), whereas blue is a dark color and has a low level of luminance (is nearer to black).

Hue is the element that distinguishes between colors (red, blue, green, etc.). White, black, and gray are not hues. Hue information is contained in the phase angle of the 3.58-MHz color carrier. The three primary video colors of red, blue, and green can be combined in such a manner as to create any hue. A phase shift through 360° produces every hue in the rainbow by changing the combination of red, blue, and green.

Saturation is the vividness of a hue, and is determined by the amount the color is diluted by white light. Saturation is often expressed in percent: 100% saturation is a hue with no white dilution, and produces a very vivid shade. Low

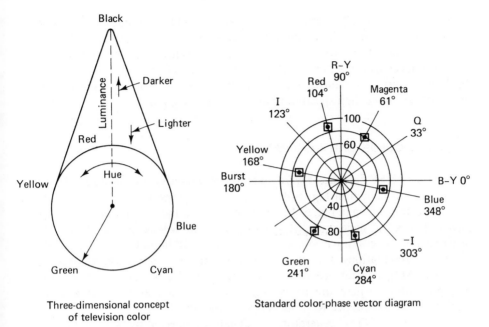

Three-dimensional concept
of television color

Standard color-phase vector diagram

Figure 6-3 Elements of a color television signal.

saturation percentages are highly diluted by white light and produce light pastel shades of the same hue. Saturation information is contained in the amplitude of the 3.58-MHz color carrier. Because the response of the human eye is not constant from hue to hue, the amplitude required for 100% saturation is not the same for all colors.

The combination of hue and saturation is known as *chroma* or chrominance. This information is usually represented by a vector diagram. Saturation is indicated by the length of the vector, and hue is indicated by the phase angle of the vector. The entire color signal representation is three-dimensional, consisting of the vector diagram for chrominance and a perpendicular plane to represent the amplitude of luminance.

6-1.6 NTSC Color Bar Signal

As discussed, the chroma amplitude required for 100% saturation of some hues is considerably greater than for other hues. Also, the luminance level for each color differs. An NTSC color bar generator produces standard EIA colors at the prescribed luminance level (brightness), chroma phase angle (hue), and chroma amplitude (saturation) set forth by the NTSC. These color signals are equivalent to the test signals used in broadcasting studios and transmitting equipment. Thus, an NTSC color bar generator makes a superior instrument for servicing and adjusting color television receivers and all types of other video equipment. A typical low-cost color generator produces all hues at the same luminance level (or with no luminance component) and the same chroma amplitude. Many hues are oversaturated. Also, the chroma phase angle is normally produced by using a carrier that is offset enough from 3.58 MHz so that a 360° phase shift occurs during each horizontal line. This produces a gated rainbow pattern rather than specific, phase-controlled colors.

6-2. NTSC COLOR BAR GENERATORS

At one time NTSC generators were quite heavy and bulky (in addition to being expensive). As shown in Fig. 6-4, today's NTSC generators are almost portable. The unit shown in Fig. 6-4 weighs about 11 pounds and requires 18 W of power. The NTSC generator differs from the rainbow types in that the NTSC instrument produces single colors, one at a time, in addition to standard NTSC color bar presentations, similar to those transmitted by television stations. A typical NTSC generator provides independent selection of fully saturated colors, plus white, where the phase angles and amplitudes are permanently established in accordance with the NTSC standards.

In a typical NTSC generator, the signals are selected by taps on a linear delay line so that no color adjustments are required. The amplitude of the color signals are also accurately set to NTSC standards, and the color reference burst is placed in its precise NTSC position, closely following the horizontal sync pulse. Thus, the color signal produced by an NTSC-type generator is exactly the same as if the signals were being produced by a television station transmitting color.

Generally, the most useful signal produced by an NTSC instrument is the standard NTSC bar pattern (as shown in Fig. 6-1) with an − IWQ signal occupying the lower quarter of the pattern, as shown in Fig. 6-5. Note that Fig. 6-5 shows both the pattern (as produced on a television screen when the NTSC

Figure 6-4 Model 1250 NTSC color bar generator. (Courtesy B&K Precision Test Instruments Product Group of Dynascan Corporation, Chicago, Illinois.)

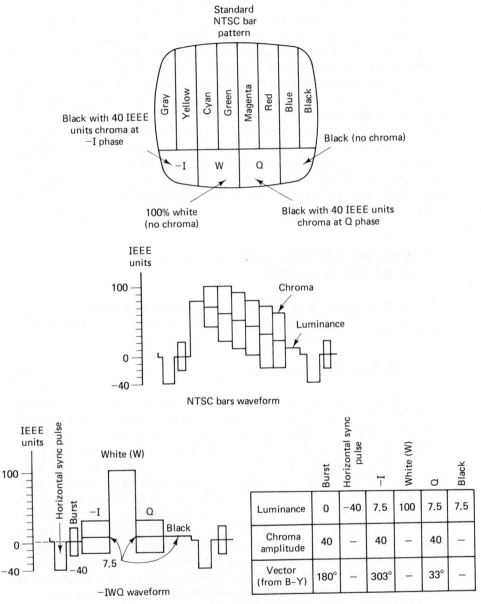

Standard
NTSC bar
pattern

Gray | Yellow | Cyan | Green | Magenta | Red | Blue | Black

Black with 40 IEEE
units chroma at
−I phase

Black (no chroma)

−I   W   Q

100% white
(no chroma)

Black with 40 IEEE units
chroma at Q phase

IEEE
units

100 —

Chroma

Luminance

0 —

−40 —

NTSC bars waveform

IEEE
units

Horizontal sync pulse

Burst

White (W)

100 —

−I   Q

Black

0 —

−40 —   −40   7.5

−IWQ waveform

	Burst	Horizontal sync pulse	−I	White (W)	Q	Black
Luminance	0	−40	7.5	100	7.5	7.5
Chroma amplitude	40	−	40	−	40	−
Vector (from B−Y)	180°	−	303°	−	33°	−

Figure 6-5 Pattern and waveforms for standard NTSC bars and −IWQ signal.

279

generator signal is applied to the antenna input), and the waveforms (as produced on an oscillscope connected to various points in the television receiver circuits) for one horizontal line of the pattern. Note that the bar's waveform is generated for the top 75% of the pattern, and the $-$IWQ waveform is on for the bottom 25% of the display. When viewing on an oscilloscope, the two waveforms may be superimposed, as discussed in Sec. 6–3.2.

The generator also produces a number of other patterns, including various convergence, linear staircase, and full-color rasters. Use of these patterns in troubleshooting is discussed in Sec. 6–3. The instrument of Fig. 6–4 generates an RF output either on channel 3 or 4, or at the standard TV IF frequency of 45.75 MHz. These carriers can be modulated by any of the patterns. In addition to the video patterns, the instrument generates a 4.5-MHz sound carrier with 1- or 3-kHz modulation, or with external modulation, a vertical or horizontal trigger output pulse, and a chroma subcarrier signal of 3.579545 MHz.

6–3. USING THE PATTERNS PRODUCED BY AN NTSC COLOR BAR GENERATOR

The following paragraphs describe how the patterns and waveforms produced by a typical NTSC color bar generator can be used for troubleshooting both television receivers and VCRs.

6–3.1 NTSC Color Bar Pattern

The pattern shown in Fig. 6–5 is the basic pattern used for most testing, troubleshooting, and adjustments in video equipment. Analysis of the pattern obtained on the screen of a television under test, or a pattern produced by playback of a VCR, can often localize a problem to a few specific circuits.

The NTSC color bar pattern is virtually a necessity for VCR service. Most VCR manufacturers specify an NTSC color bar input signal in their service literature. Adjustment procedures for VCRs are usually based on an NTSC color bar input (at the antenna or at the IF input), and observation of the waveforms (resulting from the NTSC input) shown at various points on the schematic diagram. Some VCR manufacturers provide both field and factory procedures in their literature. The field procedures eliminate all adjustments that require an NTSC color bar pattern, and thus severely restrict the amount of servicing that can be performed. For example, an NTSC color bar pattern is essential for chroma and luminance alignment.

Overall performance. An overall performance test of a VCR may be conducted by recording the NTSC color bar pattern, then playing it back on a video monitor or known-good television receiver. There should be little difference between the video played back from the VCR and an NTSC color bar pattern applied directly to the monitor or television.

Luminance and chromaproportions.   In a VCR, the luminance and chroma signals are separated during the recording process and recombined during the playback process. If luminance and chroma signals are not maintained at the proper proportions when separated, color distortion will probably result, particularly in the vividness of colors or color saturation. Waveforms may be examined throughout the VCR circuits for proper luminance-to-chroma proportions. For that reason, the waveforms shown in this section include the luminance-to-chroma proportions, expressed in IEEE units. As an example, in Figs. 6-1 and 6-5, the yellow bar occurs when the luminance is 69 units and the chroma amplitude is 62 units, at a vector of 167° (using B-Y as a 0° reference).

Luminance and chroma delay.   Another problem that may be found in VCRs is a difference in delay between the luminance signal and the chroma signal due to some circuit defect. Such a delay will cause fuzziness along the edges of the color bars. The delay problem may be more pronounced along the edges of the white bar in the −IWQ portion of the pattern.

RF and IF section performance.   The tuners and IF sections of VCRs and color television receivers are essentially the same. The RF and IF outputs of a color bar generator can be used effectively to troubleshoot these sections by comparing the patterns produced. For example, performance of a VCR or television should be nearly as good when using the RF signal on channel 3 or 4 as when applying composite video directly into the IF sections. If the display is substantially reduced when the NTSC generator output is applied through the RF (or tuner) section, as compared to when a signal is applied to the IF section, the tuner is suspect, and you have a good starting point for troubleshooting. Once you have isolated the problem to either the RF or IF sections, you can further pinpoint the trouble by injecting RF and IF signals at various points throughout the circuits, and identifying the point at which normal operation is lost.

Color adjustments.   The NTSC color bar pattern provides a standard reference for color adjustments in television receivers. The pattern contains bars of the three primary colors: red, blue, and green. These are a good reference for checking 3.58-MHz phase problems. The white, yellow, cyan, and magenta help define problems wherein the mix of colors is not in the correct proportions.

In some troubleshooting applications, it is helpful to know if the problem is chroma- or luminance-related. The generator of Fig. 6-4 is provided with a chroma-off function, which produces the bars and waveform shown in Fig. 6-6 (normal bar pattern but without color). A television can be checked by comparing the displays of Figs. 6-5 and 6-6. If the display of Fig. 6-6 is good but the

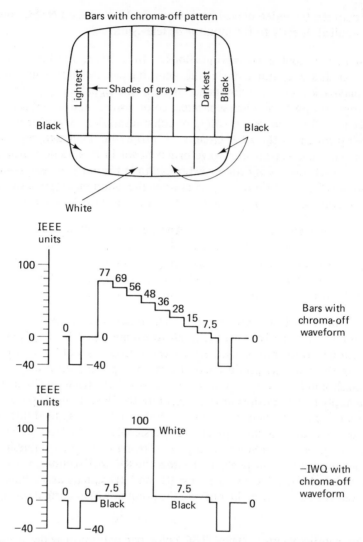

Figure 6-6   Pattern and waveforms for chroma-off function.

display of Fig. 6–5 is not good, the problem is in the chroma circuits. Note in Fig. 6–6 that the chroma-off waveform is generated for the top 75% of the pattern, and the − IWQ waveform is on for the bottom 25% of the display. When viewing on an oscilloscope, the two waveforms are superimposed.

Phase lock of the 3.58-MHz oscillator.   The 3.58-MHz oscillator within a television receiver should be locked in phase whenever the color burst is applied. This condition can be checked using the top burst-off function of the color generator. When the top burst-off function is selected, the color is unsynchronized for the top one-fourth of the pattern, but there should be no delay or color distortion where the color bars start (middle of pattern), as shown in Fig. 6–7. Note in Fig. 6–7 that the bars with burst-off waveform are generated for the top 25% of the pattern, the bars waveform (Fig. 6–5) is generated for the middle 50%, and the −IWQ waveform (Fig. 6–5) is generated for the bottom 25%. The three waveforms are superimposed on an oscilloscope.

Color killer function.   Most television receivers and VCRs have a color killer circuit that disables the color functions when there is no color burst present. This condition can be checked using the full burst-off function of the color generator. When the full burst-off function is selected, the bars should appear (in various shades of gray) but there is no color, as shown in Fig. 6–8. Note in Fig. 6–8 that the bars with burst-off waveform (Fig. 6–7) are generated for the top 75% of the pattern, and the −IWQ with full burst-off waveform is generated for the bottom 25%. The two waveforms are superimposed.

Audio or sound functions.   A 1-kHz or 3-kHz audio signal can be added to the RF output of the color generator. In a properly operating television or VCR, this audio modulation should not affect the normal display. For exam-

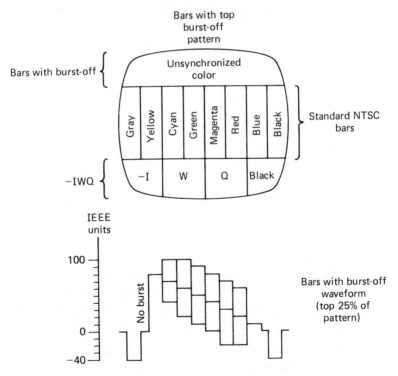

Figure 6-7 Pattern for bars with top burst off, and waveform for bars with burst off.

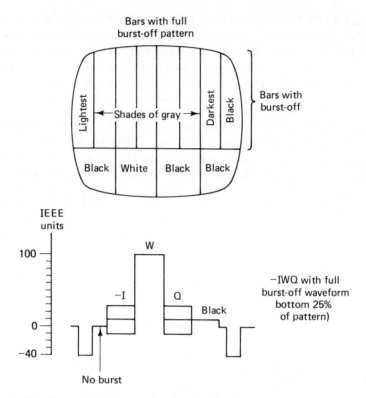

Figure 6–8 Pattern for bars with burst off, and waveform for −IWQ with full burst off.

ple, if you select the pattern of Fig. 6–5, there should be no change in the display when the audio modulation is applied or removed. If you note some interference (typically diagonal lines across the display) when audio modulation is applied, sound is leaking into the chroma circuits. Possibly one or more of the sound traps within the television or VCR are not properly adjusted. The 1-kHz and 3-kHz signals can also be used to test the sound or audio circuits of the television or VCR.

Vectorscope applications.   The NTSC color bar output (Fig. 6–5) of the generator can be used in conjunction with a vectorscope to analyze color circuits. A vectorscope measurement is often more helpful for troubleshooting than merely observing the pattern displayed on the television picture tube. The display of an NTSC vectorscope, with the NTSC color bar pattern (Fig. 6–5) applied, should be six dots located within the six boxes of the vectorscope graticule as shown in Fig. 6–3. The pattern or signal to be displayed on the vectorscope may be probed from anywhere in the composite video or 3.58-MHz color circuits. Amplitude must be initially adjusted so that the color burst dot aligns with the 75% mark on the vectorscope graticule.

If an NTSC vectorscope is not available, a good laboratory-type oscillo-scope can be substituted. The demodulated color signals directly from the red and blue guns can be used as X and Y inputs to the oscilloscope, as shown in Fig. 6-9. This connection should produce a display on the oscilloscope as shown in Fig. 6-10. The oscilloscope should be set up for vectorscope operation as follows:

1. Select the NTSC color bar pattern (Fig. 6-5), and apply the RF output of the generator to the antenna input of the television receiver, as shown in Fig. 6-9.

2. Set up the oscilloscope for X-Y operation. Adjust the position controls to center the dot on the screen with no signal input to the oscilloscope.

3. Connect both the X and Y inputs of the oscilloscope to the red gun of the television picture tube.

4. Adjust the vertical and horizontal gain of the oscilloscope to equal amounts which will move the spot to the 45° reference position (from the center) as shown in Fig. 6-10.

5. Now move the horizontal (X) input of the oscilloscope to the blue gun. Leave the vertical (Y) input connected to the red gun.

6. For a 90° picture tube, the display on the oscilloscope should be similar to that shown in Fig. 6-10. The typical 105° picture tube will produce a more elliptical display.

7. If desired, the various vectors may be selected one at a time by selecting

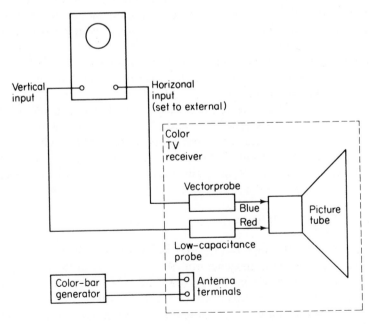

Figure 6-9 Connecting a conventional oscilloscope as a vectorscope.

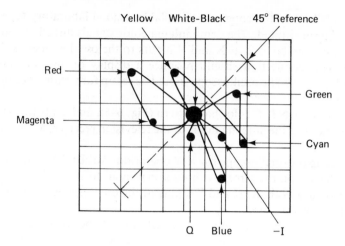

Figure 6-10 Typical vectorscope display of demodulated color bar pattern viewed on a conventional oscilloscope.

the corresponding raster color display of the color generator. Raster displays are discussed in Sec. 6–3.5.

6-3.2 – IWQ Patterns

The basic −IWQ pattern appears on a split field with the NTSC color bar pattern, with the −IWQ pattern appearing on the lower one-fourth of each field, as shown in Fig. 6–5 through 6–8. When viewing horizontal lines of video (waveforms) on an oscilloscope, both the bars and −IWQ signals are superimposed, which is very handy in many applications. The 7.5% black level (right-hand side) and the 100% white level (second from left) of the −IWQ pattern are the key luminance amplitude references. These black and white level references are used for FM deviation adjustment and black-clip and white-clip level adjustments in VCRs. The black and white levels are also used whenever luminance and chroma ratios are being adjusted, or are being checked during troubleshooting. The −I and Q signals are used primarily in video cameras and studio equipment for setting up the phase and amplitude of the −I and Q signals, and maintaining the proper relationship between the two.

6-3.3 Staircase Patterns

Figures 6–11 through 6–17 show the various staircase patterns and waveforms available with the generator of Fig. 6–4. When the pattern of Fig. 6–17 is selected, the waveforms of Fig. 6–15 and 6–16 are generated for the top 25% of the pattern, and the waveform of Fig. 6–11 is generated for the bottom 75%. Both waveforms are superimposed on an oscilloscope display.

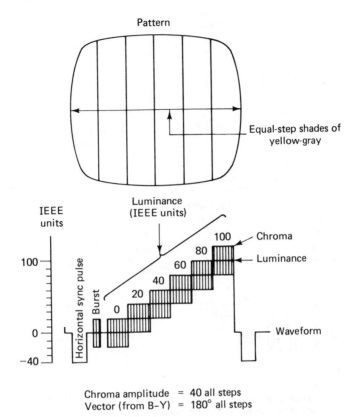

Figure 6-11   Pattern and waveform for high staircase function.

Amplifier linearity checks.   The staircase pattern contains five equal steps of increasing luminance with a constant chroma amplitude and phase. With the chroma off (Figs. 6-13 and 6-14), only the luminance steps are generated. This luminance-only pattern is valuable for checking linearity in television and VCR amplifier circuits. The amplitude of each step should be equal at the output of an amplifier or other circuit since the amplitude is equal at the input. Nonequal steps monitored at the output of a circuit represents nonlinear distortion.

Setting white-clip level in VCRs.   The staircase pattern is desirable for setting the white-clip levels in VCRs. Since the top step is 100% white level, the top step provides the correct reference for white-clip level adjustment. If the top step shows less amplitude than the other steps, this usually indicates incorrect adjustment of the white-clip level.

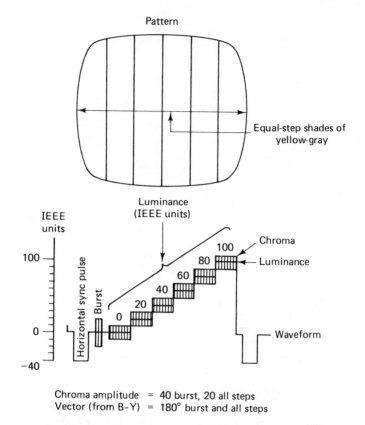

Chroma amplitude = 40 burst, 20 all steps
Vector (from B–Y) = 180° burst and all steps

Figure 6-12   Pattern and waveform for low staircase function.

Frequency equalization adjustments in VCRs.   The staircase pattern is also recommended for frequency equalization adjustment in the record amplifier of VCRs. The FM signal which carries luminance information in a VCR is shifted to a different frequency for each step of the staircase signal. However, the record current (current applied to the tape recording heads in a VCR) should remain constant across the FM frequency band. The frequency equalization adjustments of a VCR should be set so that the record current is equal for all steps of the staircase input signal.

Differential gain and differential phase checks.   The staircase patterns are most effective when checking both differential gain and differential phase. Excessive differential gain or differential phase can be the cause of color distortion in VCRs, color television receivers, and video monitors. Both conditions are checked often in studio equipment that processes video signals.

Theoretically, the chroma amplitude should not change as luminance is

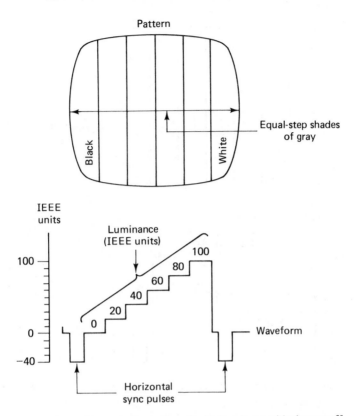

Figure 6-13   Pattern and waveform for high staircase with chroma off.

varied from 0 to 100%. Any interaction is called *differential gain*. To check for differential gain, the output of a chroma circuit is displayed on a precision waveform monitor while the staircase pattern input is applied from the generator. As the luminance signal steps from 0 to 100%, in 20% increments, any differential gain causes changes in the chroma amplitude, in synchronization with the luminance steps. Sometimes, the degree of differential gain may be affected by the peak-to-peak amplitude of the chroma signal. This characteristic can be checked by switching between the low staircase (Figs. 6-12, 6-14, and 6-16) and high staircase (Figs. 6-11, 6-13, 6-15, and 6-17) patterns. The low staircase pattern generates 20 IEEE units of chroma amplitude, whereas the high staircase generates 40 IEEE units.

Theoretically, chroma phase should not shift as luminance is varied from 0 to 100%. Any interaction is called *differential phase*. To check for differential phase, the output of the chroma circuit is displayed on an NTSC vectorscope while the staircase pattern input is applied. As the luminance signal steps from 0 to 100%, in 20% increments, any differential phase causes changes in the vector

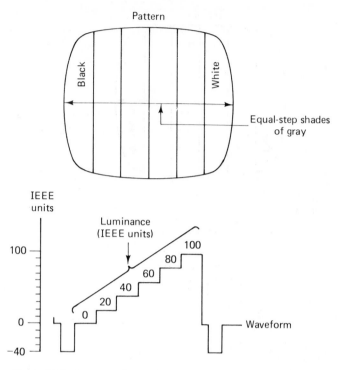

Figure 6-14 Pattern and waveform for low staircase with chroma off.

angle, in synchronization with the luminance steps. Switching between the low staircase and high staircase patterns shows whether the degree of differential phase is affected by chroma amplitude.

6-3.4 Convergence Patterns

Figures 6-18 through 6-20 show the various convergence patterns and waveforms available with the generator of Fig. 6-4. The convergence patterns are used primarily for static and dynamic convergence of color television receivers and video monitors.

Center cross.   The center cross pattern (Fig. 6-18) should intersect at the center of the screen, and there should be no tilt of the horizontal line. Improper centering indicates the need for centering adjustment, or a possible deflection circuit fault. Tilt may require repositioning of the deflection yoke for correction. This pattern also provides a good general check of vertical and horizontal sync.

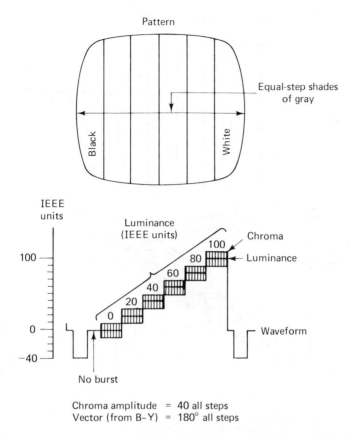

Figure 6-15   Pattern and waveform for high staircase with full burst-off function.

Dots pattern.   The dots pattern (Fig. 6-19) is used for static convergence, usually by converging the center dot of the pattern. A 19 by 15 dot pattern is generated by the instrument of Fig. 6-4. Most television receivers and some video monitors have some overscan, so that all dots are not visible, except possibly under low-voltage conditions. Some television sets have a tendency toward a greater amount of overscan than others. Typically, it is desirable to display at least a 17 by 13 dot pattern.

Crosshatch pattern.   The crosshatch pattern (Fig. 6-20) is normally preferred for dynamic convergence, although some technicians prefer the dots pattern for both static and dynamic convergence. The crosshatch pattern checks both vertical and horizontal linearity. Each square of the crosshatch pattern should be the same size, which is a convenient reference for making linearity adjustments. The crosshatch pattern is also used to check so-called pin-

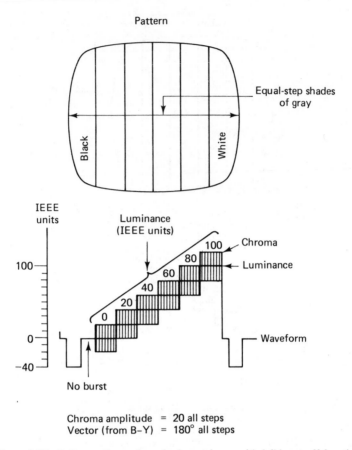

Figure 6–16   Pattern and waveform for low staircase with full burst-off function.

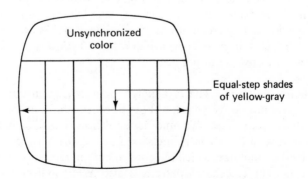

Figure 6–17   Pattern for staircase with top burst-off function.

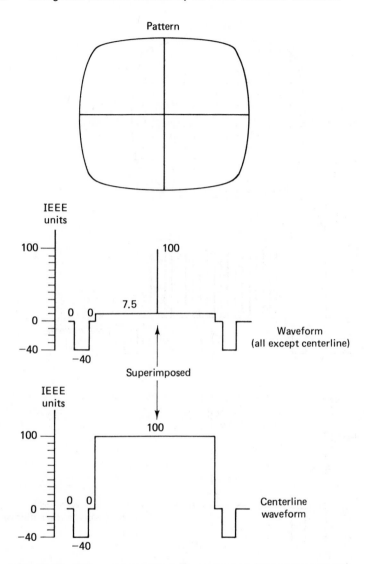

Figure 6-18 Pattern and waveforms for center cross (convergence pattern).

cushion distortion, which often appears at the outside edges of large-screen television sets as bends in the lines.

Dot hatch pattern. The dot hatch pattern (Fig. 6-21) combines the dots and crosshatch patterns for a quick overall check of static and dynamic convergence, linearity, overscan, and pincushion distortion from a single pattern.

Note that there may be a slight amount of vertical jitter when observing

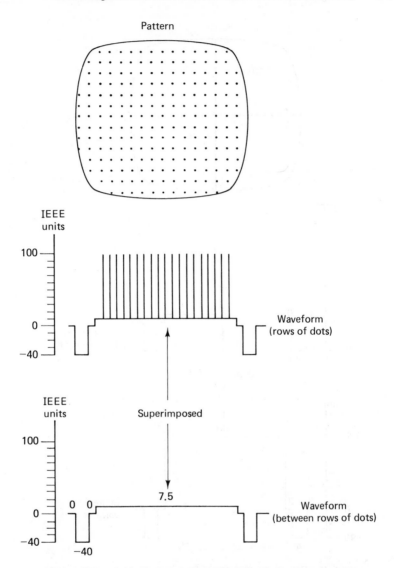

Figure 6-19   Pattern and waveforms for dots (convergence pattern).

the convergence patterns using a generator similar to the instrument in Fig. 6-4. This jitter is the result of interlaced scan and is normal for an NTSC generator (since the NTSC signal is interlaced). The jitter does not degrade the accuracy of convergence adjustments, and does not indicate a malfunction associated with vertical sync. Non-NTSC generators, such as the typical rainbow generator, do not show the jitter since their scan is not interlaced (nor do they show true NTSC signals).

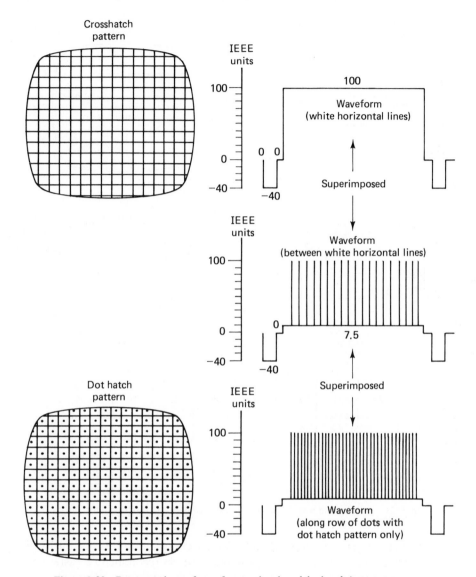

Figure 6-20   Pattern and waveforms for crosshatch and dot hatch (convergence patterns).

6-3.5 Raster Patterns

Figure 6–21 shows the raster patterns and waveforms available with the generator of Fig. 6–4. As shown, one raster pattern fills the entire screen, while the other raster pattern (with top burst off) fills the bottom 75% of the screen, with the top 25% occupied by unsynchronized color.

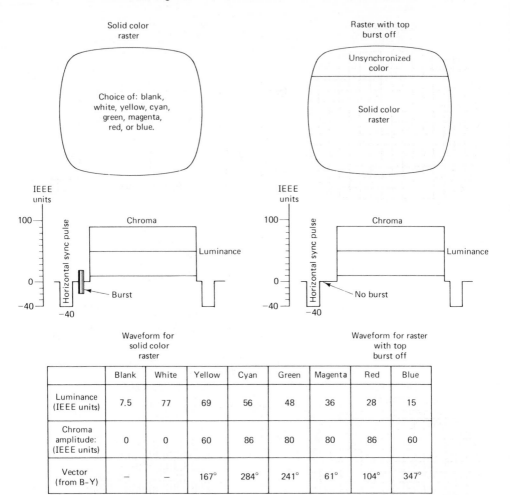

Figure 6-21 Patterns and waveforms for solid color raster and raster with top burst off.

	Blank	White	Yellow	Cyan	Green	Magenta	Red	Blue
Luminance (IEEE units)	7.5	77	69	56	48	36	28	15
Chroma amplitude: (IEEE units)	0	0	60	86	80	80	86	60
Vector (from B−Y)	−	−	167°	284°	241°	61°	104°	347°

The raster patterns are valuable for checking and adjusting purity in television receivers. Also, some VCR manufacturers recommend a raster pattern for setting the record current. Not only can the white raster be used in the standard manner, but the three separate guns of a television receiver may be individually adjusted using a continuous chroma signal of red, blue, or green. Analysis of hue and saturation problems may be simplified by analyzing each primary color, or the yellow, cyan, and magenta hues which are the equal mixture of two primary colors without the third primary color. A "black burst" test signal is generated when none of the three primary colors are selected. The luminance and chroma components for each raster color are identical to the corresponding bar from the color bar pattern, but each color can be selected individually for analysis.

6-3.6 Sync Signals

Every pattern produced by the generator of Fig. 6-4 includes NTSC sync pulses. These sync pulses are the same as those produced by a broadcast station. The sync amplitude of 40 IEEE units is often the reference against which the remainder of the luminance signal is compared. Circuits can be checked for sync clipping by observing the staircase pattern on an oscilloscope and checking whether the sync pulse amplitude remains 0.4 compared to the 100% white step, which is a reference of 1.0. AGC circuits, which respond to sync pulse amplitude, can be checked by using the composite video output of the Fig. 6-4 generator, and varying the output level. Although the overall amplitude of the video is varied, the sync pulse amplitude is a *constant percentage* of the total video amplitude.

The precise timing of the sync pulses allows proper adjustment of the servo circuits and switching circuits in VCRs. The servo circuits control the speed of the video tape in VCRs, while the switching circuits provide for switching between video heads (to allow a continuous transition from field to field).

6-3.7 Special Applications for NTSC Generators

The following are a few typical applications for NTSC generators in systems associated with television.

CCTV.   Closed-circuit television systems do not usually include a built-in NTSC color reference signal. The generator of Fig. 6-4 can be used to supply such a reference. All equipment in the CCTV system can then be adjusted to the same reference while connected into or removed from the system.

Virtually all equipment used in CCTV systems is designed around certain standard signal conditions at the input and output terminals. This allows equipment to be compatible, without modification, when interconnected as a system. Among the standard signal conditions are: positive polarity video signal (negative sync pulses), 1-V peak-to-peak signal amplitude, 75-Ω impedance, and unbalanced line (one side grounded). The generator of Fig. 6-4 also has these characteristics, which simplifies the setup for testing and troubleshooting.

CATV.   Cable television systems also use the standard signal conditions specified for CCTV at the input and output terminals. The channel 3 or 4 RF output of the Fig. 6-4 generator can also be used throughout a CATV system to test, adjust, or troubleshoot amplifiers, cables, and any other equipment.

MATV.   Master television antenna systems for hotels, motels, apartment buildings, and so on, can be checked by applying the channel 3 or 4 RF output of the Fig. 6–4 generator at the input of the network (or a branch of the network) and examining the pattern obtained on channel 3 or 4 directly from the screen of each television receiver connected to the system. To isolate problems in the distribution network from problems in an individual television set, apply the channel 3 or 4 RF output directly into the antenna terminals of the television set. If the display is good with direct connection, but not through the MATV system, you have a problem in the distribution system (or a branch of the system).

6–4. DEFINITION OF NTSC TERMS

The following are definitions of terms associated with NTSC signals.

B-Y.   A color signal at $0°$ on the standard vector diagram (Fig. 6–2). B-Y is formed from a combination of red, green, and blue ($-0.30R$ $-0.59G$ + $0.89B$). Any color on the vector diagram can be made from a combination of B-Y ($0°$) and R-Y ($90°$), or their counterparts $-$(B-Y) and $-$(R-Y) at $180°$ and $270°$, respectively.

$-$(B-Y).   A color signal at $180°$. This is also the phase of the color burst. See also "B-Y."

Back porch.   The portion of a composite video signal between the trailing edge of the horizontal sync pulse and the end of the horizontal blanking pedestal (Fig. 6–1). The color burst occurs during the back porch interval.

Bars pattern.   See "NTSC color bars."

Blanking level.   The level of the front and back porches. Zero IEEE units.

Burst.   See "color burst."

CATV.   Cable television. Also used for Community Antenna Television.

CCTV.   Closed-circuit television.

Chroma or chrominance.   The color information contained in a video signal, consisting of hue (phase angle) and saturation (amplitude) of the color carrier.

Chroma amplitude.   Amplitude of 3.58-MHz color carrier. Represents saturation.

Chroma phase angle.   Phase angle of 3.58-MHz color carrier. Represents hue.

Color bars.   See "NTSC color bars."

Color burst.   A few (8 to 10) cycles of 3.58-MHz color carrier which occurs during the back porch interval. Color burst amplitude is 40 IEEE units and phase is $180°$. The color oscillator of a color television receiver is phase-locked to the color burst.

Color carrier.   The 3.58-MHz signal which carries color information. This signal is superimposed on the luminance level. Amplitude of the color carrier represents saturation, and phase angle represents hue.

Composite video signal.   The entire video signal consisting of blanking pulses, sync pulses, color burst, and luminance and chroma information (see Fig. 6-1).

Differential gain.   A change of chroma gain as a result of varying the luminance level. The staircase patterns are most useful for measuring differential gain (which can cause saturation distortion).

Differential phase.   A phase shift of the 3.58-MHz color carrier as a result of varying the luminance level. The staircase patterns are most useful for measuring differential phase (which can cause hue distortion).

Duty cycle.   Percentage of cycle during which the pulse is working. A true square wave has a 50% duty cycle. Horizontal sync pulses have about 8% duty cycle (about 5-μs pulse width at 63.5-μs pulse repetition period).

EIA.   Electronic Industrial Association.

Equalizing pulse.   A portion of the vertical blanking interval which is made up of blanking level and six pulses (8% duty cycle at -40 IEEE units) at one-half the width of horizontal sync pulses and at twice the repetition rate (see Fig. 6-2). One equalizing pulse occurs immediately before, and another occurs immediately after, the vertical sync pulse.

Field.   One-half of a television picture (Fig. 6-2). One complete vertical scan of the picture, containing 262.5 lines. Two fields make up a complete television picture (frame). The lines of field 1 are vertically interlaced (fall halfway between the lines) with field 2 for 525 lines of resolution.

Frame.   A complete television picture, consisting of two fields. See "field."

Front porch.   Blanking level pulse at the end of the line of horizontal scan, before the horizontal sync pulse (see Fig. 6-1).

Horizontal blanking pedestal.   That portion of each line of composite video signal which blanks the picture while CRT retrace returns to the left side of the screen. Consists of front porch, horizontal sync pulse, and back porch (see Fig. 6-1).

Hue.   Distinction between colors. Red, blue, green, yellow, and so on, are hues. White, black, and gray are not considered hues (in television).

I.   Color signal at 123° (see Fig. 6-2). The I signal is also a designation for a signal containing color information in the 0- to 1.5-MHz band, representing small areas of color for picture details, before being used to modulate the 3.58-MHz color carrier. See also "$-$I" and "Q."

$-$I.   Color signal at 303° (see Fig. 6-2). An NTSC-designated phase angle reference used in studio equipment. Phase angle $-$I and Q are separated by 90°. With their counterparts I and $-$Q, a full 360° vector can be generated. In this respect, I and Q are equivalent to B-Y and R-Y.

$-$IWQ.   A pattern (Fig. 6-5) consisting of equal parts $-$I (40 units chroma at 303°), W (100% white), Q (40 units chroma at 33°), and black. The −IWQ pattern occupies the lower 75% of vertical scan below the NTSC color bar pattern.

IEEE.   Institute of Electrical and Electronic Engineers.

IEEE units.   A standard 1-V peak-to-peak composite video signal is divided into 140 equal units, scaled from −40 to +100, which are then called IEEE units. Luminance and chrominace amplitudes are measured in IEEE units. Sync pulses from 0 to −40 units, blanking level is 0, picture information spans the +7.5 setup level (black) to +100 (100% white) levels. Chroma amplitude is the peak-to-peak amplitude of the color carrier, which rides on the luminance level.

Interlace.   Vertical offset between field 1 and field 2 that causes lines of field 1 to fall between the lines of field 2. Also see "field" and "frame."

Luminance.   The amount of light intensity perceived by the eye as brightness. Luminance information is represented by the amplitude of the composite video signal.

MATV.   Master antenna television.

Monochrome.   Black-and-white television signal. Contains sync and luminance but no color burst or chroma.

NTSC.   National Television Systems Committee. Established the color television standards now in use in the United States and many other nations of the world.

NTSC color bars.   A pattern (Fig. 6–5) consisting of eight equal parts. Colors are white (75%), black (7.5 setup level), 75% saturated pure colors red, green, and blue, and 75% saturated hues of yellow, cyan, and magenta (mixtures of two colors in a 1:1 ratio without a third color).

Q.   Color signal at 33° (see Fig. 6–2). The Q signal is also a designation for a signal containing color information in the 0- to 0.5-MHz band, representing large areas of color, before being used to modulate the 3.58-MHz color carrier. See also "I" and "−I."

−Q.   A color signal at 213° (see Fig. 6–2). See also "Q" and "I."

R-Y.   A color signal at 90° on the standard vector diagram (see Fig. 6–2). R-Y is formed from a combination of red, green, and blue chrominance (+0.70R −0.59G −0.11B). See also "B-Y."

−(R-Y).   A color signal at 270° (see Fig. 6–2). See also "R-Y" and "B-Y."

Saturation.   Vividness of color. Degree to which a color is not diluted by white light. Highly saturated color is very vivid. The same hue becomes a pastel shade when diluted by white light. Saturation is represented by chroma amplitude and is measured in IEEE units. The number of IEEE units for fully saturated color varies from hue to hue.

Setup.   The separation between blanking and black reference levels (see Fig. 6–1). The NTSC standard setup level is 7.5 units.

Staircase.   A pattern (Figs. 6–11 and 6–12) consisting of five equal steps of luminance level and a constant-amplitude, color burst chroma signal. Chroma amplitude is selectable at 20 IEEE units (low staircase) or 40 IEEE units (high staircase). The staircase pattern is useful for checking linearity of luminance and chroma gain, differential gain, and differential phase.

VCR.   Video cassette recorder. Used interchangeably with VTR (video tape recorder) in some literature. However, a VTR may or may not use cassette tape. Studio VTRs use reel-to-reel tape.

Vertical blanking interval.   That portion at the beginning of each field of composite video signal which blanks the picture while the CRT retrace returns to the top of the screen. The equalizing pulses and vertical sync pulse are generated within this interval (see Fig. 6-2).

Vertical sync pulse.   A portion of the vertical blanking interval which is made up of blanking level and six pulses (92% duty cycle at −40 IEEE units) at twice the horizontal sync pulse repetition rate. Synchronizes vertical scan of television receiver to composite video signal. Starts each frame at the same vertical position. (Alternating fields are offset one-half line to achieve interlaced scan.)

Y signal.   The black-and-white portion of a video signal. Same as luminance.

Index